INTRODUCTION TO CHEMICAL KINETICS

INTRODUCTION TO
Chemical Kinetics

GORDON B. SKINNER
Department of Chemistry
Wright State University
Dayton, Ohio

ACADEMIC PRESS New York and London 1974

A Subsidiary of Harcourt Brace Jovanovich, Publishers

ACADEMIC PRESS, INC.
111 Fifth Avenue, New York, New York 10003

United Kingdom Edition published by
ACADEMIC PRESS, INC. (LONDON) LTD.
24/28 Oval Road, London NW1

Library of Congress Cataloging in Publication Data

Skinner, Gordon.
Introduction to chemical kinetics.

Includes bibliographies.
1. Chemical reaction, Rate of. I. Title.
QD502.S55 541′.39 73-18979
ISBN 0-12-647850-3

PRINTED IN THE UNITED STATES OF AMERICA

Contents

Preface

Like many other textbooks, this one started as a set of lecture notes for a course. Over a period of years, parts of it were typed up and distributed in class to supplement the textbooks then in use. Finally, it seemed only natural to organize and add to the material to produce a textbook.

Our students have not, I suspect, been too different from those in comparable first-year graduate courses in kinetics at other medium-sized universities. All of our students have taken physical chemistry as undergraduates, but few remember it very well. Hardly any planned to be physical chemists, let alone kineticists. Most have done their dissertations in polymer, analytical, or inorganic chemistry.

Accordingly, the text includes brief reviews of several thermodynamic and structural concepts in places where they are used, and a longer section on the calculation of partition functions and their use in calculating equilibrium constants—methods essential in the use of the Activated Complex Theory. The book has a strongly practical slant. Throughout, careful attention is given to units, and many numerical problems are given for students to solve, a good many based on recent journal publications. Students who have solved these problems should have little difficulty coping with the considerable variety of units used to express kinetic data in the literature.

Although most research papers proceed from experimental to results to discussion, this text follows the reverse sequence, experience having shown that this order makes sense pedagogically. Students have no difficulty supposing that kinetic data can be measured somehow, and once they have seen what sort of kinetic data a chemist needs to know, they are able to appreciate the methods that have been developed. Sometimes I have lectured to the class on experimental methods, but more often I have had each student write a paper on a method that interested him, then present a summary to the class.

My colleagues and I agreed that a modern kinetics course simply had to involve the use of a computer. Many kinds of kinetic calculations are impractical without one. Fortunately, many of our students have taken a computer course, while we have found that the others could begin writing useful programs with the help of a 30-page set of notes (written in English, not Computerese) and a firm push to overcome the high activation energy associated with the first reaction of student with computer. Several of the problems in the text, while solvable by a slide rule or desk calculator, are suitable for illustrating the value of a computer in kinetic calculations, while detailed computer solutions of three typical problems are given in the text.

Several topics currently of great interest to kineticists are omitted or treated only briefly. These include, for example, molecular beam experiments, and calculations of potential energy surfaces and molecular trajectories. While this work is starting to provide important insights into the nature of chemical reactions, practical applications are limited so far. What we have done is to point out that these ideas are being developed, and are worth watching.

Because this text starts at the beginning of kinetics, it would be suitable as an undergraduate text in a physical chemistry course sequence that includes one quarter of quantum mechanics, one of thermodynamics, and one of kinetics. Discussions in the text on the nature of liquids and solids (as these affect kinetic behavior) and the wide variety of problems will help to broaden the students' thinking into related areas of physical chemistry—something to be desired in a three-part course.

I would like to acknowledge the suggestion of my wife, Marjorie, that I write the text, as well as her patience while I completed it, since it took at least a factor of two longer than I estimated in my first burst of enthusiasm. The Chemistry 751 class of the Winter Quarter, 1973, contributed by using the first draft and discovering a number of unclear passages. Acknowledgments should also go to Mrs. Jane Wyen, who typed the manuscript, and to the staff of Academic Press, who were most efficient and helpful.

1 The Nature of Chemical Kinetics

Among fields of study there are some that have close relationships to our daily lives and are, therefore, obviously important. Specialists in these fields can see developments occurring all the time as they go about their business. For example, the political scientist, the sociologist, and the economist can read about events related to these subjects every time they pick up a newspaper. Similarly, a chemist specializing in the control of atmospheric pollution can check on the success of his work whenever he steps out of doors in the morning and sniffs the air.

The relationship of chemical kinetics to daily life is not so obvious, yet the perceptive kineticist can see his science in action in many places. If one is driving a car and sees a traffic light change, the time needed to respond to the signal is determined by the speeds of a sequence of relatively fast biochemical reactions. The internal combustion engine works because, among other things, the rate of combustion of hydrocarbon fuels is slow at low temperatures, so that the fuel can safely be mixed with air, but fast at high temperatures, so that a few milliseconds are enough time for fairly complete combustion. Steel, aluminum, and most plastics are all thermodynamically unstable toward oxidation by the atmosphere, but kinetically steel tends to oxidize more rapidly. Steel is so desirable from mechanical and economic points of view that an enormous and fairly successful technology has developed to protect steel surfaces from rusting, but even so one of the conspicuous marks of our technology is the large amount of rusting steel junk we leave lying around. There are, besides, other relationships. I think that a kineticist who has worked out the details of a complex process involving, perhaps, 20 different simultaneous reactions may have a special appreciation of the difficulty of solving a social problem in which business, labor, consumer, governmental, legal, and political interests are all interacting.

Probably, though, kinetics enters the life of a typical chemist when he asks, How fast does this reaction go under certain conditions, and how can I change the rate? Usually, before he gets very far into the kinetics of a reaction, the chemist has looked into two other basic chemical aspects: the *nature* of the substances involved and the extent to which the reaction goes at *equilibrium*. The latter aspect, of course, is thermodynamic; and a kineticist must realize that he is, to a large extent, a slave to thermodynamics. He can hope only to understand and perhaps control the rate at which the reaction goes toward equilibrium.

Within these limitations, though, a chemist has considerable freedom to use his knowledge of kinetics to practical advantage. For example, if one is interested in making *p*-dichlorobenzene (a quite effective moth repellent) by the chlorination of benzene, he is faced with the problem that, once one chlorine has been added, there is an almost equal a priori likelihood of the second adding ortho or para. As it happens, *o*-dichlorobenzene is not much use for making mothballs or anything else; all the ortho product that is made is both a waste of starting material and a serious disposal problem. Accordingly, much research has gone into the development of catalysts that speed up the formation of *p*-dichlorobenzene but not ortho, and considerable success has been achieved.

From the scientific point of view (need I say it?) kinetics can be a fascinating field of study. The direction of much current research is toward the most complete understanding possible of what happens as two molecules approach, react, and gradually (over a period of perhaps 10^{-12} sec) change into reaction products. This is the very essence of chemistry! The area of kinetics is one of the least understood, even at a much less sophisticated level than that. As an example, only within the last ten years have the actual chemical processes (involving H, OH, O, and HO_2 radicals) in the hydrogen–oxygen reaction been understood, and the rates of some of these subreactions are still known only approximately. Similarly, the rate of the reaction of hydrogen ions with hydroxide ions to form water has been measured only recently, and at the present time the kinetics of many comparably common chemical reactions are unknown or incompletely known.

Because a chemist must use thermodynamics and a knowledge of molecular structure in obtaining, understanding, or using kinetic data, I have included some material from these fields in this book. Of course, to talk about the chemistry of real molecules one must know some organic and inorganic chemistry as well; but since the purpose of the book is to describe the methods and principles of kinetics, I have generally used examples that will be familiar to students of varied backgrounds. Once he has mastered these ideas, each student should be able to apply them in his own specialized area.

2 How Kinetic Results Are Expressed

In this chapter we will concentrate on the notation used in describing the results of experiments in chemical kinetics, without going into detail as to how the results are obtained. As with most subjects, conventional ways of presenting data have been developed and are fairly generally followed, but there are variations that reflect both the personal preferences of the researchers and the different levels of the total topic of kinetics that are being described.

2.1 REACTION RATES

It is usual to consider the rate of a reaction to be the change in concentration per unit time of any reaction product for which the stoichiometric coefficient is 1. This is a satisfactory definition as long as the volume of the reacting system remains constant, which is usually true or nearly so—for example, for reactions in dilute solution or for gas reactions carried out in closed containers. When the volume changes (Section 2.8), we can think of the rate of reaction as the rate of change of concentration that would occur due to reaction if the volume did not change.

Probably the most common units of concentration are moles per liter (mole ℓ^{-1}), but such units as moles per cubic centimeter (mole cm^{-3}) and molecules per cubic centimeter (molecules cm^{-3}) are fairly often used. So far the International System of Units (SI) (1) has been little used in reporting kinetic results, but one would expect that, in the future, units of moles per cubic meter (mole m^{-3}) will be widely used.[1] The unit of time is usually the second, though larger time units are sometimes used for slower reactions.

For example, if we study the reaction

$$2\ HI(g) \rightarrow H_2(g) + I_2(g)$$

at constant volume and obtain a rate of change of I_2 concentration of 1.5×10^{-5} mole ℓ^{-1} sec^{-1}, then that is the rate of the reaction. From stoichiometry we also know that

$$\frac{d[H_2]}{dt} = 1.5 \times 10^{-5} \quad \text{mole } \ell^{-1} \text{ sec}^{-1}$$

and

$$\frac{d[HI]}{dt} = -3.0 \times 10^{-5} \quad \text{mole } \ell^{-1} \text{ sec}^{-1}$$

For any reaction product, the rate of change of concentration is equal to the stoichiometric coefficient times the rate of reaction; while for the reactants, the rate of change of concentration is equal to the negative of the stoichiometric coefficient times the rate of reaction.

In writing chemical equations a method that lends itself well to computer programming has recently become popular. One rearranges the equation so that the substances are all listed on one side, with negative stoichiometric coefficients for the reactants; for example,

$$H_2(g) + I_2(g) - 2\ HI(g) = 0$$

In generalized form, one would write the equation as

$$\sum_i \nu_i A_i = 0 \tag{2.1}$$

where A_i is the symbol of a product or reactant, and ν_i is its stoichiometric coefficient, ν_i being positive for products and negative for reactants. With this redefinition of the stoichiometric coefficients, if we symbolize the rate of a reaction by R, then

$$\frac{d[A_i]}{dt} = \nu_i R \tag{2.2}$$

[1] The author has noted with some annoyance that "mol" has been recommended as the abbreviation for "mole" in the SI system. Since some authors use "mol." as an abbreviation for "molecule," confusion can readily arise, particularly since many reactions have rate constants in the range of 10^{12} in mole cm^{-3} units and, correspondingly, in the range of 10^{-12} in molecule cm^{-3} units. We will abbreviate neither "mole" nor "molecule" in this book.

which will apply to both products and reactants at constant volume (provided, of course, that they are not participating in any other reactions at the same time).

From the above statements, it is clear that in principle one can measure the rate of a reaction by measuring the rate of change of concentration of any reactant or product. From an analytical point of view, one substance might be much easier to measure than another. Besides this, one must be on the lookout for side reactions. For example, if a product reacted further by a second reaction, the rate of its concentration change would not be a suitable quantity for measuring the rate of the first reaction.

2.2 CLASSIFICATION OF CHEMICAL REACTIONS

Reactions may be categorized in many ways: organic and inorganic, for example, or as occurring in the gaseous, liquid, or solid phase. However, there is one less-common method of classification that is of importance in kinetics. Professor H. S. Johnston put it this way:

> I find it very convenient to think of gas phase chemical kinetics at three different "levels of abstraction," that is, degrees of averaging: (1) over-all chemical reactions, (2) elementary chemical reactions, and (3) elementary chemical–physical reactions. An ideal elementary chemical–physical reaction involves the transition of reactants each in a definite quantum state to products each in a definite quantum state: such transitions are purely mechanical processes and independent of temperature. An elementary chemical reaction consists of one type of elementary chemical–physical reaction averaged over a steady (not necessarily equilibrium) distribution of states, such a distribution depending on only a small number of macroscopic variables. Elementary chemical reaction rates depend on temperature, concentration, and perhaps other variables. Practical, over-all reactions typically involve a set of elementary chemical reactions, occurring in series or parallel or both; the set of elementary steps is often referred to as the "mechanism." (2)

An elementary reaction, then, is one that occurs exactly as written. If we say the reaction

$$OH^- + CH_3Br \rightarrow CH_3OH + Br^-$$

is an elementary one, we mean that an OH^- ion collides with a CH_3Br

molecule, producing a short-lived intermediate from which a CH_3OH molecule and a Br^- ion result directly. Many gas-phase reactions involving atoms and free radicals are elementary, such as

$$Br(g) + H_2(g) \rightarrow HBr(g) + H(g)$$

and

$$H(g) + C_2H_6(g) \rightarrow C_2H_5(g) + H_2(g)$$

A number of decomposition and rearrangement reactions, such as

$$\underset{\text{methyl isocyanide}}{CH_3NC(g)} \rightarrow \underset{\text{acetonitrile}}{CH_3CN(g)}$$

are very likely elementary, as are some of the very rapid inorganic solution reactions such as

$$NH_4OH \rightarrow NH_4^+ + OH^-$$

although in these a proper statement of the reaction process should include a description of the solvation of the reactants and products.

The conclusion that a reaction is elementary usually follows a serious attempt to prove that it is not. The process is analogous to deciding whether a given substance is an element, and the possibility always exists that future research will show that a certain reaction is not really elementary. The classic example is the reaction

$$H_2(g) + I_2(g) \rightarrow 2\,HI(g)$$

which from the time it was first studied by Bodenstein (3) in 1894 gave every evidence of being elementary and was widely used as an example in textbooks. However, in the 1960s Sullivan (4) showed beyond doubt that the mechanism includes a pair of elementary reactions:

$$I_2(g) \rightarrow 2\,I(g)$$

$$2\,I(g) + H_2(g) \rightarrow 2\,HI(g)$$

At the time of writing of this book, a lively discussion is in progress (5) as to whether the reaction

$$H_2(g) + D_2(g) \rightarrow 2\,HD(g)$$

is elementary.

Elementary reactions are sometimes classified according to their *molecularity*, that is, the number of reactant molecules participating in the chemical change that is occurring. For example, the decompositions of CH_3NC and I_2 are *unimolecular*, while the reactions of H with C_2H_6 and of I with H_2 are *bimolecular*. If a reaction is not elementary, then it is not appropriate to talk about its molecularity. Because it is not easy (as we have suggested above) to find out whether a reaction is elementary, and because the role of energy exchange among molecules (a basically physical process) has not been clearly understood until recent years, many incorrect assignments of molecularity have been made in the past. It is now perfectly clear that the molecularity of a reaction and its reaction order (an experimental reaction property which is discussed in the next section) are two completely different and distinguishable entities even though, in some cases, they may have the same numerical value. The definitions of molecularity in the first sentence of this paragraph and of reaction order in Section 2.3 should lead to a clear distinction between the two.

The study of elementary chemical–physical reactions, as Johnston terms them, is still in its infancy. The apparatus needed to produce monoenergetic reactants and analyze for the products is very sophisticated, and the number of kinetic systems studies so far is small. One current result of this work is a growing appreciation among kineticists of the role of the energy distribution among and within reacting molecules on the observed rates of reaction.

2.3 REACTION ORDER

The rate of a reaction is usually found to depend on the concentration of some of the reactants and is often influenced by the presence of other substances deliberately or accidentally added to the reaction mixture. At a given temperature, and perhaps within a limited range of concentrations, one can write a *rate law* for a reaction of the form

$$\text{rate} = k \prod_i [A_i]^{\alpha_i}[X_j]^{\beta_j} \tag{2.3}$$

where the A_i are the reactants, the X_j are substances that are not reactants but do influence the rate, the α's and β's are coefficients that are not necessarily related to the stoichiometric coefficients ν, and k is a *rate constant*. For a long time it was thought that the α's and β's would be integers, but it is now clear that they need not be.

Two kinds of reaction order are commonly defined. The *overall order* of a reaction is the sum of all the α's and β's in the rate law expression. The overall order tells us how the rate responds to changes in the absolute concentration at constant relative concentration, such as one would produce by changing the pressure in a gaseous system or diluting a liquid system with an inert solvent. The *order with respect* to A_i is α_i, and similarly the order with respect to X_j is β_j. These individual orders not only tell us how sensitive the system is to changes in the concentration of each species, but may also suggest the chemical mechanism of the reaction.

For many elementary reactions, the rate law coefficients are equal to the stoichiometric coefficients. This is true for some of the elementary reactions given above, provided the concentrations are moderate. That is, for

$$OH^- + CH_3Br \rightarrow CH_3OH + Br^-$$

the rate law is

$$\text{rate} = k[OH^-][CH_3Br]$$

in dilute solutions. Since the reaction is elementary, the rate depends on the number of collisions between the reactants, which in turn depends on the product of the concentrations. In a different solvent of different viscosity the number of collisions per second at given concentrations would change, the effect being accounted for by a change in k. In concentrated solutions, the number of collisions per unit time will no longer be directly proportional to the concentrations, so the above simple rate law will not hold. For reactions between ions, the long-range nature of the interactions shows (as we will see later in more detail) that concentrations need not be very high before nonidealities become noticeable.

For the reaction

$$Br(g) + H_2(g) \rightarrow HBr(g) + H(g)$$

the rate law again has a simple form

$$\text{rate} = k[Br][H_2]$$

since in a gas (except at very high pressures) the number of collisions per second is proportional to this product of the concentrations. For the isocyanide rearrangement,

$$CH_3NC \rightarrow CH_3CN$$

the rate law is

$$\text{rate} = k_1[CH_3NC]$$

at high pressures, but changes to

$$\text{rate} = k_2[CH_3NC]^2$$

at low pressures with a transition region between, where the order varies from 1 to 2. The explanation of this[2] is that, while the reaction is chemically elementary, it is complex from a physical (energetic) point of view, and at low pressures the rate is governed by the "physical" elementary reaction

$$2\,CH_3NC \rightarrow CH_3NC^* + CH_3NC$$

[2] Originally suggested by F. A. Lindemann (6). A further discussion of this topic is given in Chapter 4.

where CH_3NC^* is a molecule with enough energy to react. This problem does not arise for solution reactions since the solvent molecules are a large and readily accessible source of energy.

In general, one can expect that the rate law for an elementary reaction will be a simple one, and that the α values may well be equal to the stoichiometric coefficients. Experience will be a guide as to which types of reaction follow this generalization most closely. The rates will, of course, have these simple forms only at the beginning, when no products are present. Once appreciable quantities of products form, the reverse reaction will begin to occur, so that, for example, the rate of the Br + H_2 reaction will become

$$\text{rate} = k[\text{Br}][\text{H}_2] - k_{\text{R}}[\text{HBr}][\text{H}]$$

where k_R is the rate constant of the reverse reaction. The presence of the reverse reaction leads to an observed rate expression that is not usually a product of powers of concentrations, so to obtain the rate law for a given reaction (in one direction) it is simplest to study the rate near the beginning of the reaction or, to be more general, under conditions far from equilibrium. Unless we make a point of mentioning otherwise, the "rate" of a reaction will imply the forward rate. A rate constant will *always* refer to reaction in one indicated direction.

Let us look at some examples of rate laws for nonelementary reactions. For the decomposition of HI to H_2 and I_2, Bodenstein (3) found experimentally that the rate law is

$$\text{rate} = k[\text{HI}]^2$$

the value 2 being determined to within about ± 0.1. The reaction, then, is of the second order both overall and with respect to HI. If the rate is 1.5×10^{-9} mole ℓ^{-1} sec^{-1} at an HI concentration of 2×10^{-3} mole ℓ^{-1}, then we can calculate k:

$$1.5 \times 10^{-9} \quad \text{mole } \ell^{-1} \text{ sec}^{-1} = k(2 \times 10^{-3} \quad \text{mole}^{-1})^2$$

$$k = 3.75 \times 10^{-4} \quad \text{mole}^{-1}\ \ell \text{ sec}^{-1}$$

The units of k are chosen to make the rate equation balance dimensionally.

For the reverse reaction

$$H_2 + I_2 \rightarrow 2\ HI$$

the rate law is

$$\text{rate} = k_{\text{R}}[\text{H}_2][\text{I}_2]$$

The overall order is again 2, while the orders with respect to H_2 and I_2 are each 1. As indicated above, the integral reaction orders suggested that this reaction is elementary, but further research has shown it is not.

For the reaction

$$H_2 + Br_2 \rightarrow 2\ HBr$$

also studied by Bodenstein (7), the initial rate is given by the equation

$$\text{rate} = k[H_2][Br]^{1/2}$$

so that the overall order is $1\frac{1}{2}$, the order with respect to H_2 is 1 and that with respect to Br_2 is $\frac{1}{2}$. This different rate law immediately suggested that the reaction mechanism is complex, and in time a mechanism of several steps (Section 4.2) involving H and Br atoms was deduced.

In the above examples the reaction orders are integers or common fractions since it happens that the rate laws are the result of chemical mechanisms that lead to such simple orders. A different situation has recently been reported by Hartig *et al.* (8); they found that in the temperature range 2000–2500°K and pressure range 5–250 atm, the rate of decomposition of methane in a dilute methane–argon mixture could be expressed as

$$\text{rate} = k[CH_4][Ar]^{0.4}$$

Here the argon is acting as an energy transfer medium, and there are good reasons to believe that its order is both temperature and pressure dependent. Accordingly, the statement that the argon order is 0.4 not only states the fact observed in the range studied, but suggests that a different order might be found at other conditions. Since chemists are usually looking (and hoping) for simplicity, it is customary when a reaction order has been found experimentally to be close to an integer (as that of methane was in this study) to round it off to the integer, whereas a value such as 0.4 cannot be rounded off very well and is allowed to stand as it is.

It can be seen that, if the overall order of a reaction is n, then the units of the rate constant are given by

$$\frac{(\text{concentration})\,(\text{time})^{-1}}{(\text{concentration})^{n}} \qquad \text{or} \qquad (\text{concentration})^{1-n}\ \text{time}^{-1}$$

For the methane example, $n = 1.4$ and the units of k would be $(\text{mole}\ \ell^{-1})^{-0.4}$ sec^{-1} or $\text{mole}^{-0.4}\ \ell^{0.4}\ \text{sec}^{-1}$.

2.4 THERMODYNAMIC CONSISTENCY OF RATE LAWS

For the elementary reaction

$$H(g) + Br_2(g) \rightarrow HBr(g) + Br(g)$$

we have written that the rate is

$$\text{rate} = k[\mathrm{H}][\mathrm{Br_2}] - k_\mathrm{R}[\mathrm{HBr}][\mathrm{Br}]$$

where k and k_R are the forward and reverse rate constants, respectively. At equilibrium, the rate becomes 0, and the equation can be rearranged to give

$$\frac{k}{k_\mathrm{R}} = \frac{[\mathrm{HBr}][\mathrm{Br}]}{[\mathrm{H}][\mathrm{Br_2}]} = K$$

where K is the thermodynamic equilibrium constant for the reaction if we assume ideal behavior, so that the concentrations are equal to the activities—a good assumption for gases but not so good for liquids. In general, for any elementary reaction, the numerical ratio of the forward and reverse rate constants must be the equilibrium constant, and there must also be dimensional equivalence.

Before going on to consider complex reactions, we should pause to consider the question of concentration units. For the $H + Br_2$ reaction, it is clear that K is dimensionless, and the k and k_R must have the same units ($\text{mole}^{-1}\ \ell\ \text{sec}^{-1}$, for example). For any elementary reaction for which the number of moles of substance does not change, the units of the forward and reverse rate constants will be the same, and there will be no difficulty as to the units.

If we consider, however, an elementary reaction such as

$$\mathrm{H_2(g) + Ar(g) \rightarrow 2\,H(g) + Ar(g)}$$

for which the forward reaction has been found to be second order (first order with respect to each of H^2 and Ar) and to have a rate constant of $2.2 \times 10^4\ \text{mole}^{-1}\ \ell\ \text{sec}^{-1}$ at 3000°K (9), we do have a little problem in calculating the reverse reaction rate since most tables of gas-phase equilibria (such as the JANAF tables[3]) list K_p, the equilibrium constant in terms of pressure, while we need K_c, the equilibrium constant in terms of concentration. The key is found in the ideal gas law

$$PV = nRT$$

$$\text{concentration} = \frac{n}{V} = \frac{P}{RT} \tag{2.4}$$

[3] The JANAF tables (10) are an excellent compilation of thermochemical data, compiled at Dow Chemical Company under the guidance of a Joint Army, Navy, and Air Force committee. It is available from the United States Government Superintendent of Documents, Washington, D.C. as NSRDS-NBS-37.

For this example, then,

$$K_c = \frac{[\mathrm{H}]^2}{[\mathrm{H}_2]} = \frac{(P_{\mathrm{H}}/RT)^2}{P_{\mathrm{H}_2}/RT} = \frac{1}{RT}\frac{(P_{\mathrm{H}})^2}{P_{\mathrm{H}_2}} = \frac{1}{RT}K_p$$

If we insert the numerical values $T = 3000°\mathrm{K}$, $R = 0.082\ \ell$ atm mole^{-1} deg^{-1}, $K_p = 2.5 \times 10^{-2}$ atm, we get[4]

$$K_c = 1.02 \times 10^{-4} \quad \text{mole } \ell^{-1}$$

Accordingly, if $k/k_{\mathrm{R}} = K_c$

$$k_{\mathrm{R}} = \frac{k}{K_c} = 2.2 \times 10^8 \quad \text{mole}^2\ \ell^{-2}\ \text{sec}^{-1}$$

The reverse reaction is third order and, since it is elementary, we can write the rate law as

$$\text{rate} = k_{\mathrm{R}}[\mathrm{H}]^2[\mathrm{Ar}]$$

For liquid solution reactions, equilibrium constants are likely to be found in mole ℓ^{-1} units, so this conversion will not be necessary.

For nonelementary reactions, the above relationships are not applicable. It is still true, of course, that at equilibrium the rates of forward and reverse reactions should be the same, but here equilibrium defines the concentrations of one or more intermediate species as well as the requirement that the concentrations of the reactants and products should be in the ratio defined by the equilibrium constant. If, say, the reactants present at equilibrium were suddenly removed, then the concentrations of the intermediates might well change to new values, and the rate of the forward reaction would take on a value different from that at equilibrium. The change would not, of course, be predictable without detailed knowledge of the reaction mechanism.

In general, it is hazardous to extrapolate kinetic data on a nonelementary reaction beyond the experimental range. The rate law may change, and the rate "constant" and its temperature dependence may change, both in ways

[4] Strictly speaking, equilibrium constants are dimensionless, and instead of actual concentrations or pressures, we put into the above equation ratios of actual concentrations or pressures to the standard state concentrations or pressures. In converting from K_p to K_c, the $1/RT$ term is also taken to be dimensionless, being merely the ratio of the concentrations in the two standard states of 1 atm and 1 mole l^{-1}. However, if one uses this convention, then all rate constants and rates must have the same units (inverse seconds, for example) and we obtain the rates by multiplying rate constants by ratios of actual concentrations to standard (unit) concentrations. Since attaching concentration units to equilibrium and rate constants is such a useful method of keeping track of reaction orders, we will continue to do so, as have most practical kineticists.

that are unpredictable in the absence of knowledge about the mechanism. This difficulty explains, in large part, why a knowledge of reaction mechanism has great practical as well as strictly scientific importance.

2.5 REACTION RATES AS DIFFERENTIAL QUANTITIES

So far the rate of a reaction has been discussed as if it were an observable quantity, but in most instances it is not. Usually a chemist will measure the concentration of a substance as a function of time, and deduce from this the rate during the time. From the definitions of rate laws we have used it is clear that since the concentrations are changing with time, the rates are also. To obtain rate constants from such measurements of concentration, one has two basic choices, both of which have been used: (a) carry out the reaction to a small extent, so that the concentrations of reactants change little during the experiment; or (b) integrate the rate law expression so that the rate constant may be determined from values of concentrations obtained at known times. Since analytical requirements frequently dictate the second approach, kineticists have obtained integrated forms of most imaginable rate laws. We will look at a few of the simpler ones in detail and give references to additional forms.

2.6 FIRST-ORDER REACTIONS

The rate for a first-order reaction can be written

$$\text{rate} = k[\text{A}]$$

where A is usually one of the reactants, possibly the only one. If A has a (positive) stoichiometric coefficient ν_{A} in the equation for the reaction, then *at constant volume*

$$\text{rate} = -\frac{1}{\nu_{\text{A}}}\frac{d[\text{A}]}{dt} \tag{2.5}$$

and we may make this substitution to obtain

$$d[\text{A}] = -k\nu_{\text{A}}[\text{A}]\,dt \tag{2.6}$$

$$\Delta[\text{A}] = -k\nu_{\text{A}}[\text{A}]\,\Delta t \tag{2.7}$$

so that

$$k = -\frac{\Delta[\text{A}]}{\nu_{\text{A}}[\text{A}]\,\Delta t} \tag{2.8}$$

Of course, $\Delta[A]$ will be negative, so a positive value of k will be found. This approach is very successful if one of the products of reaction is easily identified and measured in small quantities, so that a $\Delta[A]$ of perhaps 1% of [A] can be accurately measured.

On the other hand, if we plan to measure [A] only, it is much more satisfactory to carry out the reaction to a substantial extent and to integrate the equation for comparison with experiment. We separate the variables

$$\frac{d[A]}{[A]} = -k\nu_A \, dt \tag{2.9}$$

and integrate from an initial concentration of $[A]_0$ at $t = 0$ to a concentration of [A] at time t:

$$\ln [A] - \ln [A]_0 = -k\nu_A t \tag{2.10}$$

This equation indicates that a straight-line graph will be obtained for a first-order reaction if ln [A] is plotted as a function of t. The intercept will be $\ln [A]_0$ and the slope $-k\nu_A$, so that k can be obtained. Also, the equation can be solved for k

$$k = \frac{1}{\nu_A t} \ln \frac{[A]_0}{[A]} \tag{2.11}$$

and if the equation is first order, k's calculated from various pairs of t and [A] should be equal. Finally, once the value of k and the fact of first-order behavior have been established, the equation can be rearranged to give [A] as a function of time and k:

$$[A] = [A]_0 \exp(-k\nu_A t) \tag{2.12}$$

First-order behavior has been found for many chemical reactions, as well as for the important physical process of radioactive decay. For radioactivity in particular, it is the custom to express the speed of the reaction in terms of the half-life $t_{1/2}$, the time needed for the concentration of reactant to drop to half its initial value. Since $\nu_A = 1$ for nuclear processes of this type,

$$[A]_{1/2}/[A]_0 = \exp(-kt_{1/2}) = \tfrac{1}{2} = e^{-0.693} \tag{2.13}$$

$$kt_{1/2} = 0.693, \qquad t_{1/2} = 0.693/k \tag{2.14}$$

It is characteristic of first-order reactions that the time for half of the initial concentration of reactant to disappear is independent of the value of the initial concentration. For other reaction orders, this is not so; the half-life depends in some way on the reactant concentration. Determination of how the half-life depends on the concentration is one way of determining reaction order.

A good example of a first-order reaction is the gas-phase isomerization of *cis*- to *trans*-2-butene, a reaction that has been found by Rabinovitch and Michel (11) to be first order at pressures above a few Torr in the range 700–800°K. At 724°K the rate constant for the reaction

$$cis\text{-2-butene} \rightarrow trans\text{-2-butene}$$

is $1.9 \times 10^{-5}\ \text{sec}^{-1}$. Let us calculate the concentration of reactants and products as a function of time, assuming an initial pressure of 10 Torr in a reactor of constant volume.

$$[cis\text{-2-butene}]_0 = \frac{P}{RT} = \frac{10/760}{0.082 \times 742} = 2.16 \times 10^{-4} \quad \text{mole}\ \ell^{-1}$$

Considering only the forward reaction,

$$[cis\text{-2-butene}] = 2.16 \times 10^{-4} \exp(-1.9 \times 10^{-5}\, t)$$

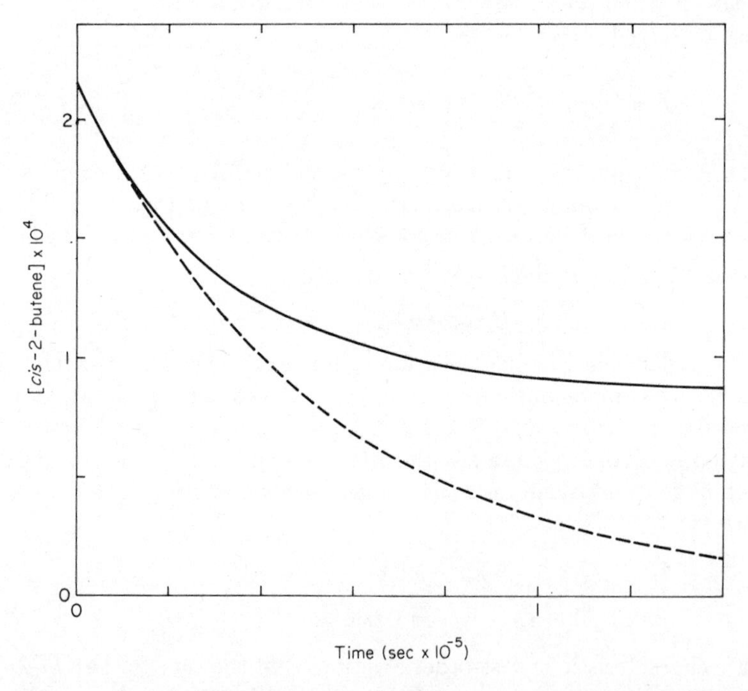

Fig. 2.1 A typical first-order reaction (cis–trans isomerization of *cis*-2-butene). Dashed curve, forward reaction only; solid curve, both forward and reverse reactions.

Let us resist, for the moment, the temptation to plot a straight line and make a direct plot of concentration versus time (the dashed line of Fig. 2.1). This exponential curve is typical of first-order reactions with no reverse reaction.

In fact, the equilibrium constant of this reaction is not far from 1. From the American Petroleum Institute tables (12) one can calculate the value to be 1.53, indicating that the reaction will proceed to somewhat more than 50%. Specifically, if [*cis*-2-butene] = x at equilibrium, then

$$K = \frac{2.16 \times 10^{-4} - x}{x} = 1.53$$

from which $x = 0.85 \times 10^{-4}$ mole ℓ^{-1}, while [*trans*-2-butene] = 1.31×10^{-4} mole ℓ^{-1}. We may also calculate k_R for the reaction since the process is though to be elementary. Thus $k_R = (1.9 \times 10^{-5})/1.53 = 1.24 \times 10^{-5}$ sec^{-1}.

How does the concentration actually vary as it approaches equilibrium? If we continue to symbolize [*cis*-2-butene] by x, and use a subscript 0 for the initial concentration,

$$\frac{dx}{dt} = -kx + k_R(x_0 - x) = k_R x_0 - x(k + k_R)$$

On separating variables and integrating, we obtain

$$\int_0^x \frac{dx}{k_R x_0 - x(k + k_R)} = \int_0^t dt$$

$$x = x_0 \left[\frac{k \exp(-(k + k_R)t) + k_R}{k + k_R} \right] \tag{2.15}$$

where x_0 = [*cis*-2-butene] at $t = 0$. The value of x has been calculated and plotted as the solid line of Fig. 2.1.

It is a good thing to check such a kinetic equation for "reasonableness" by putting in limiting values of the variables and seeing how the equation simplifies. We can check it in several ways:

If $k_R \to 0$, $x \to x_0 e^{-kt}$.
If $k \to 0$, $x \to x_0$ (no change occurs).
If $t \to 0$, $x \to x_0$.
If $t \to \infty$, $x \to x_0 k_R/(k + k_R)$ (the equilibrium value).

Since all of these checks are reasonable, we can have some confidence that a gross mathematical error has not been made in the derivation.

2.7 SECOND-ORDER REACTIONS

Rate equations fall into three groups, depending on whether the rate law depends on the second power of one reactant:

$$\text{rate} = k[\mathrm{A}]^2 \tag{2.16}$$

or on the first power of two reactants:

$$\text{rate} = k[\mathrm{A}][\mathrm{B}] \tag{2.17}$$

There is, of course, no reason that a rate should not depend on other powers of reactant concentrations, such as $\frac{3}{2}$ or $\frac{1}{2}$, but such combinations are uncommon. The second equation above breaks down mathematically into two cases, depending on whether or not $[\mathrm{A}]_0$ and $[\mathrm{B}]_0$ are equal.

In the first instance above if we again set

$$\text{rate} = -\frac{1}{\nu_\mathrm{A}}\frac{d[\mathrm{A}]}{dt} \tag{2.18}$$

then

$$d[\mathrm{A}] = -k\nu_\mathrm{A}[\mathrm{A}]^2\,dt \tag{2.19}$$

Initially, then,

$$\Delta[\mathrm{A}] = -k\nu_\mathrm{A}[\mathrm{A}]_0^2\,\Delta t \tag{2.20}$$

and

$$k = -\frac{\Delta[\mathrm{A}]}{\nu_\mathrm{A}[\mathrm{A}]_0^2\,\Delta t} \tag{2.21}$$

To integrate over a time interval, we separate the variables to get

$$\int_{[\mathrm{A}]_0}^{[\mathrm{A}]}\frac{d[\mathrm{A}]}{[\mathrm{A}]^2} = -k\nu_\mathrm{A}\int_0^t dt \tag{2.22}$$

$$-\frac{1}{[\mathrm{A}]}\Bigg]_{[\mathrm{A}]_0}^{[\mathrm{A}]} = -k\nu_\mathrm{A}t \tag{2.23}$$

$$\frac{1}{[\mathrm{A}]} - \frac{1}{[\mathrm{A}]_0} = k\nu_\mathrm{A}t \tag{2.24}$$

Accordingly, a plot of $1/[\mathrm{A}]$ versus time will yield a straight line with y intercept $1/[\mathrm{A}]_0$, while the slope gives the value of k.

For the second type of rate law,

$$\text{rate} = k[A][B],$$

general solutions would involve the use of two stoichiometric coefficients ν_A and ν_B. While these solutions can be worked out, the mathematics become involved and the principles can be better illustrated if we solve the special case of $\nu_A = \nu_B = 1$. If so, then if $[A]_0 = [B]_0$, $[A] = [B]$ in general, and the equation reduces to the first type, so we get

$$\frac{1}{[A]} - \frac{1}{[A]_0} = kt \tag{2.25}$$

$$\frac{1}{[B]} - \frac{1}{[B]_0} = kt \tag{2.26}$$

If $[A]_0$ is not equal to $[B]_0$, then let

$$[A]_0 = a_0, \qquad [B]_0 = b_0$$

$$[A] = a_0 - x \qquad [B] = b_0 - x$$

$$\frac{d[A]}{dt} = -\frac{dx}{dt}$$

$$\text{rate} = \frac{dx}{dt} = k(a_0 - x)(b_0 - x) \tag{2.27}$$

Rate constants can be found from initial rates by the equation

$$k = \frac{\Delta x}{a_0 b_0 \, \Delta t} \tag{2.28}$$

To integrate over time, we separate the variables and use the method of partial fractions[5] to obtain

$$\frac{1}{b_0 - a_0} \ln \frac{a_0(b_0 - x)}{b_0(a_0 - x)} = kt \tag{2.29}$$

[5] When the variables are separated, we obtain

$$\int_{x=0}^{x} \frac{dx}{(a_0 - x)(b_0 - x)} = \int_0^t k \, dt$$

The left-hand equation cannot be integrated directly, but the method of partial fractions (see a standard mathematics text such as "Calculus," 2d Ed., by Smith *et al.* (13)) states that

$$\frac{1}{(a_0 - x)(b_0 - x)} = \frac{c_1}{a_0 - x} + \frac{c_2}{b_0 - x}$$

where c_1 and c_2 are constants. The values of c_1 and c_2 may be found by considering how we would reconstitute the right-hand fractions over a common denominator. The numer-

This equation will not work if $b_0 = a_0$, but this situation is already covered by Eq. (2.25).

2.8 EFFECTS OF VOLUME CHANGES

As mentioned, the integrated equations of Sections 2.6 and 2.7 apply only to systems at constant volume. When the volume of the reaction system changes as the reaction proceeds, the relationships between rates of change of concentration and reaction rates become more complex.[6] Let us look at an example.

As we will see in more detail in Chapter 4, the reaction

$$C_2H_6(g) \rightarrow C_2H_4(g) + H_2(g)$$

can be first order under some conditions. Suppose we plan to study it in the first-order region, at constant pressure rather than constant volume. Clearly, the volume will increase as the reaction proceeds, being twice as great at the end of the reaction as at the beginning, so the concentration of ethane will decrease because of the increasing volume as well as because of the decreasing quantity of ethane, as the reaction proceeds. Accordingly, we would write an analogue of Eq. (2.6), taking ν_A to be 1 in this case

$$d[A] = -k[A]\,dt - [A]\frac{dV}{V} \tag{2.30}$$

---

ator would be $c_1(b_0 - x) + c_2(a_0 - x)$ and this must be equal to 1 for all values of x. This being so, the sum of the constant terms must be 1 and the sum of the coefficients of x must be zero:

$$c_1 b_0 + c_2 a_0 = 1, \qquad -c_1 - c_2 = 0$$

These two linear equations in c_1 and c_2 can be solved simultaneously to give

$$c_1 = \frac{1}{b_0 - a_0} \quad \text{and} \quad c_2 = \frac{-1}{b_0 - a_0}$$

so the original equation can be written in the integrable form

$$\frac{1}{b_0 - a_0}\int_0^x \frac{dx}{a_0 - x} - \frac{1}{b_0 - a_0}\int_0^x \frac{dx}{b_0 - x} = k\int_0^t dt$$

This procedure can easily be expanded to include a larger number of linear terms in the denominator of the original fraction.

[6] A good discussion is given by Canagaratna (14), with some more worked-out examples. He recommends a redefinition of the "rate of a chemical reaction" so that the qualification about constant volume in the usual definition (Section 2.1) can be removed. At present we prefer to stick with the usual definition, while remaining alert to its limitation.

where V is the volume of the system, dV the change in volume occurring in the time increment dt, and A stands for ethane. This general equation for a first-order reaction with volume change may be integrated in our example if we work out the relationship between the volume and the concentration of ethane.

The relationship can be obtained by noting that if x is the fraction of ethane that has reacted,

$$V = V_0(1 + x) \tag{2.31}$$

where V_0 is the initial volume, so that

$$dV = V_0\,dx \tag{2.32}$$

Also,

$$[\mathrm{A}] = \frac{1 - x}{1 + x}[\mathrm{A}]_0 \tag{2.33}$$

from which

$$\frac{dV}{V} = \frac{dx}{1 + x} = \frac{-d[\mathrm{A}]}{[\mathrm{A}]_0 + [\mathrm{A}]} \tag{2.34}$$

Substitution of Eq. (2.34) into Eq. (2.30), separation of variables, and integration leads to

$$\ln \frac{2[\mathrm{A}]}{[\mathrm{A}]_0 + [\mathrm{A}]} = -kt \tag{2.35}$$

which may be compared with Eq. (2.10). Comparison shows that the concentration of ethane drops more rapidly in the constant-pressure situation, but the fraction of the original ethane remaining at a given time is the same for both cases.

Of course, Eq. (2.35) is not very general: a different relationship between the volume and the extent of reaction would lead to a different equation. We can also visualize examples where the volume depends on the time, as for a reaction in a gas that is being compressed by a piston. For most situations, it will be possible to write a differential equation, like (2.30), but if this equation is complex the integration to obtain concentrations as functions of time may have to be done by numerical methods, probably on a computer. An example of a problem solved by numerical integration is given in the next section.

2.9 GENERALIZED EMPIRICAL RATE LAWS

Laidler (15) has listed many integrated rate equations for various reaction orders and stoichiometric coefficients. The use of such equations in the interpretation of kinetic results has diminished since it has been demonstrated that few reactions having complicated rate laws are elementary, and therefore their rate laws will not remain constant, or have integral or simple fractional coefficients, over extensive ranges of composition or temperature. This being so, a more realistic approach to such complicated processes is to seek an empirical rate law by a numerical integration carried out on a computer. Since modern computers can do this sort of calculation very quickly, one simply tries out a series of rate laws, choosing the one that best reproduces the experimental data.

As an example, suppose that between substances A and B there is a reaction for which the stoichiometry has been found to be

$$A + 2B \rightarrow \text{products}$$

Suppose also that the reverse reaction rate has been found to be negligible, and that there is a substance C that is suspected of being a catalyst to the reaction. With these assumptions, at a given temperature and volume the rate law will be of the form

$$\text{rate} = -\frac{d[A]}{dt} = -\frac{1}{2}\frac{d[B]}{dt} = k[A]^{\alpha_1}[B]^{\alpha_2}[C]_0^{\alpha_3} \tag{2.36}$$

where the subscript of $[C]_0$ reminds us that the concentration of the catalyst does not change during the reaction.

To determine the values of α_1, α_2, α_3, and k, it would be desirable to do a series of experiments with various starting concentrations, in each experiment measuring the concentration (of either A or B—in this example A) at several times. A set of experimental results could be as shown in Table 2.1.

Approximate values of the reaction orders may be obtained by comparing the rates of reaction during the initial time period for pairs of experiments in which only one initial concentration varied. For example, for the first two solutions only the initial concentration of B was changed. The initial rates are proportional to the amounts of A used up, and the concentrations of B in each experiment may be taken as the average of the initial values and the values at the end of the first time increment. Thus,

$$\frac{0.030}{0.020} \approx \left(\frac{0.170}{0.080}\right)^{\alpha_2}, \qquad \alpha_2 \approx 0.54$$

Similarly, $\alpha_1 \approx 0.98$ and $\alpha_3 \approx 0.39$. The computer was asked to make calculations at intervals of 0.1 in α over a range of 0.4 above and below these approximate values, a total of $9 \times 9 \times 9 = 729$ sets of calculations. The method was briefly as follows. Since the stoichiometry is

$$A + 2B \rightarrow \text{products},$$

Table 2.1 Simulated Experimental Data for Reaction A + 2B → Products with a Potential Catalyst C

Component	Initial Concentrations (mole l^{-1})			
	Solution 1	2	3	4
A	.100	.100	.100	.200
B	.200	.100	.200	.200
C	.100	.100	.200	.100

Time (sec)	Concentrations of A (mole l^{-1})			
	Solution 1	2	3	4
0	.100	.100	.100	.200
2000	.070	.080	.061	.141
4000	.052	.070	.046	.118
6000	.043	.064	.034	.109
8000	.038	.057	.029	.103

let x be the change in concentration of A, $2x$ the change in concentration of B, and a_0, b_0, c_0 the initial concentrations of A, B, and C. Then

$$\frac{dx}{(a_0 - x)^{\alpha_1}(b_0 - 2x)^{\alpha_2} c_0^{\alpha_3}} = k\, dt \tag{2.37}$$

For the computer to integrate the left-hand side numerically, dx is changed to Δx, a small finite change in concentration, which we set at the value 0.001. The right-hand side integrates directly, so that with a slight rearrangement one has

$$k = \frac{\Delta x}{t c_0^{\alpha_3}} \sum \frac{1}{(a_0 - x)^{\alpha_1}(b_0 - 2x)^{\alpha_2}} \tag{2.38}$$

where x has values from 0 up to the observed value. The number of terms in the sum is simply the change in x during the time interval divided by Δx. For example, for the first solution, 30 terms were used for the first 2000 sec, 18 for the next, and so on.[7] The program obtains a rate constant for each experimental point, calculates the average k and the standard deviation σ (as a fraction of k) (see Fig. 2.2 and Table 2.2).

[7] The accuracy of the calculation was increased slightly by setting the initial concentrations of A to $\Delta x/2$, and of B to Δx, less than the actual values. This meant that during each increment the average, rather than the initial, value of concentration was used.

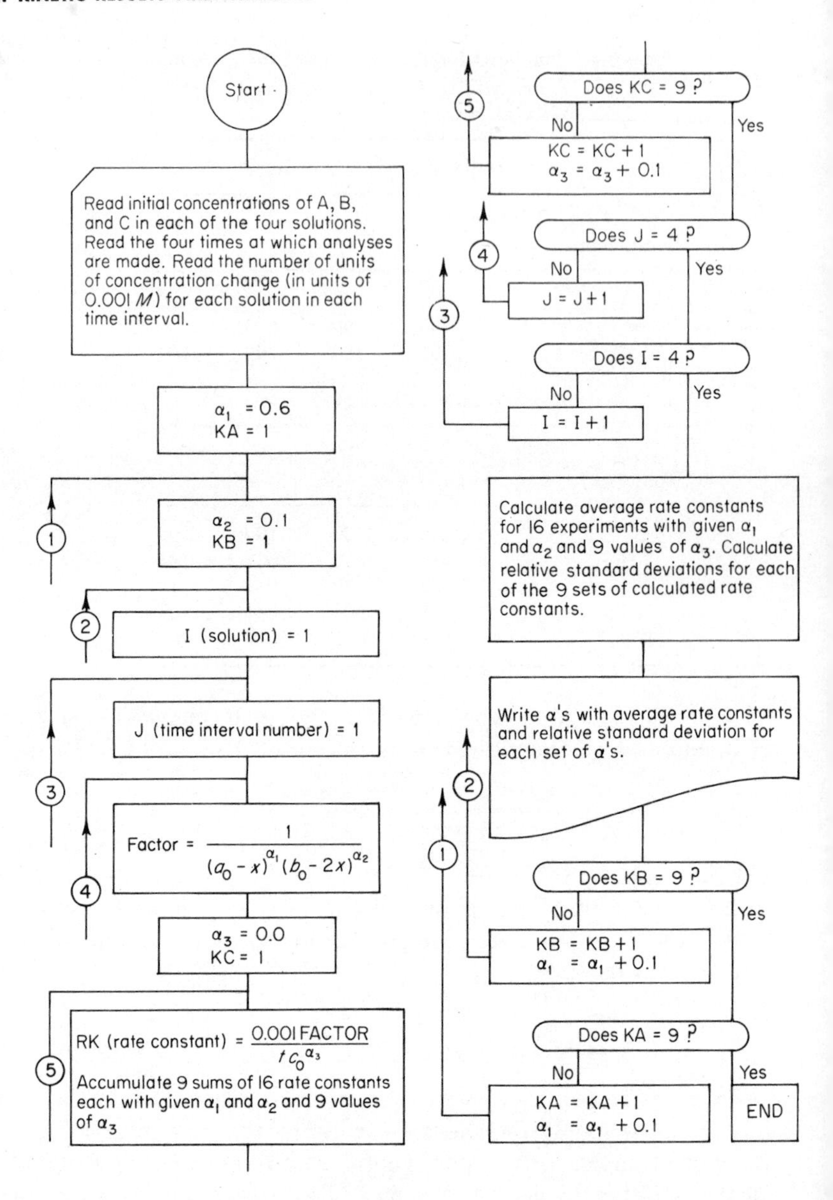

Fig. 2.2 Flow chart for computer program for calculating the rate of a reaction.

Table 2.2 A Simple Computer Program for Calculating the Rate Law of a Reaction

```
C      PROGRAM KOCAL
       DIMENSION CONCA(4),CONCB(4),CONCC(4),TIME(4),NINC(4,4),TOK(9),
      1 SUMSQ(9),RK(4,4,9),AVK(9)
       WRITE(6,200)
  200  FORMAT(1H1,5X,'RATE LAW FOR A + 2B REACTION',//,3X,' ALPHA 1.  ALP
      1HA 2     ALPHA 3        AVERAGE K          STD DEV ',/)
       READ(5,100)(CONCA(I),CONCB(I),CONCC(I),I=1,4)
  100  FORMAT(3F10.4)
       READ(5,101)(TIME(J),J=1,4)
  101  FORMAT(4F10.0)
       READ(5,102)((NINC(I,J),J=1,4),I=1,4)
  102  FORMAT(4I5)
       ALPA = .6
       DO 10 KA=1,9
       ALPB = .1
       DO 20 KB=1,9
       DO 30 KC=1,9
       TOK(KC) = 0.
       SUMSQ(KC) = 0.
   30  CONTINUE
       DO 40 I=1,4
       FACTOR = 0.
       CONA = CONCA(I)
       CONB = CONCB(I)
       DO 50 J=1,4
       NIN = NINC(I,J)
       DO 60 NI=1,NIN
       FACTOR = FACTOR + 1./(CONA**ALPA*CONB**ALPB)
       CONA = CONA - .001
       CONB = CONB - .002
   60  CONTINUE
       ALPC = 0.
       DO 70 KC=1,9
       RK(I,J,KC) = .001*FACTOR/(CONCC(I)**ALPC*TIME(J))
       TOK(KC) = TOK(KC) + RK(I,J,KC)
       ALPC = ALPC + .1
   70  CONTINUE
   50  CONTINUE
   40  CONTINUE
       ALPC = 0.
       DO 80 KC=1,9
       AVK(KC) = TOK(KC)/16.
       DO 90 I=1,4
       DO 110 J=1,4
       SUMSQ(KC)=SUMSQ(KC)+(RK(I,J,KC)-AVK(KC))*(RK(I,J,KC)-AVK(KC))
  110  CONTINUE
   90  CONTINUE
       SD = SQRT(SUMSQ(KC)/16.)/AVK(KC)
       WRITE(6,201)ALPA,ALPB,ALPC,AVK(KC),SD
  201  FORMAT(1H ,2X,3F10.1,2E15.5)
       ALPC = ALPC + .1
   80  CONTINUE
       ALPB = ALPB + .1
   20  CONTINUE
       ALPA = ALPA + .1
   10  CONTINUE
       STOP
       END
```

For these data, the lowest σ/k was 0.0162 for $\alpha_1 = 1.2$, $\alpha_2 = 0.8$, and $\alpha_3 = 0.5$, while the next lowest was 0.0294 for $\alpha_1 = 1.2$, $\alpha_2 = 0.8$, and $\alpha_3 = 0.6$; the first set is decidedly the best. The calculation was then repeated with α's at intervals of 0.02 above and below the best values, which gave a lowest σ/k of 0.0155 for $\alpha_1 = 1.20$, $\alpha_2 = 0.78$, and $\alpha_3 = 0.52$. The average experimental uncertainty in the α's (the variation in α that caused σ to double) was 0.07, varying slightly from one to another. The rate constant for the best values of α's was $(4.06 \pm 0.07) \times 10^{-3}$ mole$^{-1.5}$ $\ell^{1.5}$ sec^{-1}, the experimental uncertainty being calculated from σ/k.

The details of this calculation have been given at length because few simple examples of the use of this approach are available in the literature. Those who have done much computer programming will realize that the program could readily be improved by having the computer vary the α's systematically so as to minimize σ; this would eliminate many of the calculations that were done, although the cost of these calculations was only a few dollars. The advent of the computer means that a kineticist need no longer adopt an oversimplified model of his experimental system just because a more realistic one would involve a lot of arithmetic. If a chemist working with kinetics does not learn to be a proficient computer programmer, he should at least learn what kinds of problems can be effectively solved by using a computer, and should be sure to include at least one programmer among his friends.

2.10 EFFECT OF TEMPERATURE ON REACTION RATES

From the time of the earliest kinetic studies it has been known that reaction rates may be temperature dependent. In 1889 Arrhenius (16) proposed the equation

$$k = Ae^{-E/RT} \tag{2.39}$$

where A and E are (to a first approximation) temperature independent. This equation has been found to apply to a wide range of kinetic results, and while we have reasons to believe that both A and E do depend somewhat on temperature, experimental data are seldom accurate enough to show this effect, and the Arrhenius equation in its simple form is adequate to express most data. Frequently the equation is written in logarithmic form

$$\ln k = \ln A - \frac{E}{RT} \tag{2.40}$$

or

$$\log k = \log A - \frac{E}{2.3RT} \tag{2.41}$$

so that a graph of $\log k$ versus $1/T$ yields a straight line of slope $-E/2.3R$, while the intercept is $\log A$. A typical set of data is shown in Fig. 2.3.

It is common to speak of E as the "activation energy" and of A as the "preexponential factor." The first term is based on the idea, proposed by Arrhenius, that E is a measure of the amount of energy needed to bring about reaction. Experimental activation energies range in magnitude from near zero to over 100 kcal. Figure 2.4 shows for a generalized reaction the relationship between the thermodynamic energy change and the kinetic activation energy.

In the figure, E_F is the Arrhenius activation energy for the forward reaction, E_R that for the reverse reaction, and ΔE the internal energy change in the reaction, taken here to be negative, so that $E_F - E_R \approx \Delta E$. The relationship is approximate because, as we will see below, one must care-

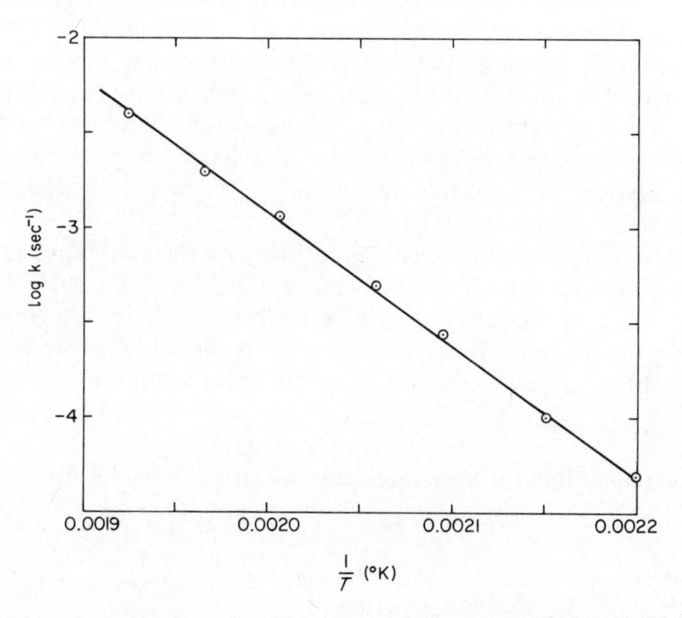

Fig. 2.3 Arrhenius plot for the reaction $CHF_2CH_2SiF_2CH_3 \rightarrow CH_2CHF + CH_3SiF_3$ in the gas phase (17).

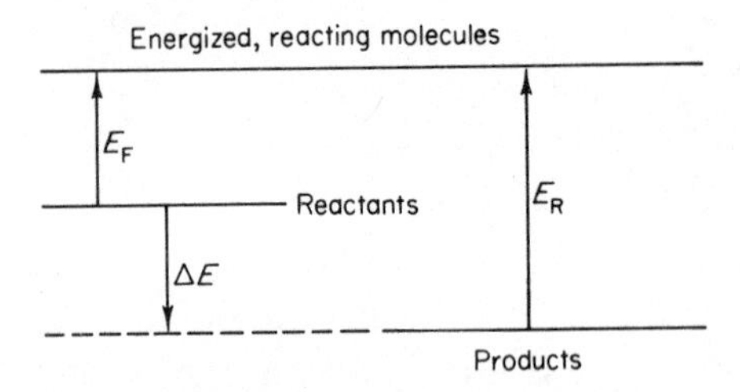

Fig. 2.4 Simplified view of the relationship between the activation energies for the forward (E_F) and reverse (E_R) reactions and the thermochemical energy change (ΔE) in a reaction.

fully specify the nature of the reaction and the experimental conditions before precise relationships can be written.

The Arrhenius equation, then, implies that reacting molecules are in an unusual, highly energized state that few molecules occupy at any one time. Moreover, molecules reacting in either direction go through the same energized state while reacting. With a few well-recognized exceptions, these statements have been found to be generally true.

The calculation of activation energies from other available information about the reacting molecules is, at this writing, one of the major unsolved problems of kinetics. It is receiving much attention from theoretical chemists, but the methods are difficult and time-consuming, and so far have been limited, with only partial success, to very simple gas reactions. It seems clear that we must depend on experimentally determined activation energies for some time to come. On the other hand, several methods, some quite successful, are available for calculating the preexponential factor A. We will look into these in Chapter 3.

At this point, though, let us look further into the correlation of experimental kinetic data with thermodynamic information about the reaction. The approach of Section 2.5 may be extended to include temperature as a variable, but again the discussion is limited to elementary reactions. If we express the equilibrium constant in an Arrhenius form:

$$K_c = C\,e^{-D/RT} \tag{2.42}$$

and the forward and reverse rate constants in the same form:

$$k = A\exp(-E/RT), \qquad k_R = A_R\exp(-E_R/RT) \tag{2.43}$$

then since

$$\frac{k}{k_R} = K_c \tag{2.44}$$

we obtain

$$\frac{A \exp(-E/RT)}{A_{\mathrm{R}} \exp(-E_{\mathrm{R}}/RT)} = C\, e^{-D/RT} \tag{2.45}$$

If this relationship is to hold over a range of temperatures, then

$$\frac{A}{A_{\mathrm{R}}} = C \tag{2.46}$$

and

$$E - E_{\mathrm{R}} = D \tag{2.47}$$

One can use these relationships to calculate one rate constant if the other one and the equilibrium constant are known, or to check kinetic data for consistency if rate data in both directions can be obtained. In instances that are comparatively rare (since the acquisition of thermodynamic data usually precedes that of kinetic) one can get a value of a hard-to-measure equilibrium constant from two rate constants.

For elementary gaseous reactions for which there is no change in the number of moles, or for elementary reactions of liquids, solids, or solutions for which volume changes on reaction are small enough to be negligible, one may write

$$K_c = K_p = \exp\left(\frac{-\Delta G^\circ}{RT}\right) = \exp\left(-\frac{\Delta H^\circ - T\,\Delta S^\circ}{RT}\right)$$

$$= \exp\left(\frac{\Delta S^\circ}{R}\right) \exp\left(\frac{-\Delta H^\circ}{RT}\right) \tag{2.48}$$

When this expression for the equilibrium constant is set equal to the ratio of the rate constants

$$\frac{A \exp(-E/RT)}{A_{\mathrm{R}} \exp(-E_{\mathrm{R}}/RT)} = \exp\left(\frac{\Delta S^\circ}{R}\right) \exp\left(\frac{-\Delta H^\circ}{RT}\right) \tag{2.49}$$

one can identify

$$\frac{A}{A_{\mathrm{R}}} = \exp\left(\frac{\Delta S^\circ}{R}\right) \tag{2.50}$$

and

$$E - E_{\mathrm{R}} = \Delta H^\circ \tag{2.51}$$

The separation of variables implies that ΔS° and ΔH° do not change over the temperature range. This assumption is usually valid over a range for which k varies by a factor of 100 or so.

If there is a change in the number of moles Δn in a gaseous reaction, then

$$K_c = \left(\frac{1}{RT}\right)^{\Delta n} K_p = \left(\frac{1}{RT}\right)^{\Delta n} \exp\left(\frac{\Delta S^\circ}{R}\right) \exp\left(\frac{-\Delta H^\circ}{RT}\right) \tag{2.52}$$

so

$$\frac{A \exp(-E/RT)}{A_R \exp(-E_R/RT)} = \left(\frac{1}{RT}\right)^{\Delta n} \exp(\Delta S^\circ/R) \exp(-\Delta H^\circ/RT) \tag{2.53}$$

If we take logarithms,

$$\ln A - \frac{E}{RT} - \ln A_R + \frac{E_R}{RT} = -\Delta n \ln(RT) + \frac{\Delta S^\circ}{R} - \frac{\Delta H^\circ}{RT} \tag{2.54}$$

and differentiate with respect to T

$$\frac{E}{RT^2} - \frac{E_R}{RT^2} = -\frac{\Delta n}{T} + \frac{\Delta H^\circ}{RT^2} \tag{2.55}$$

we can then multiply by RT^2 to obtain

$$E - E_R = \Delta H^\circ - \Delta n\, RT \tag{2.56}$$

If we now divide this equation by RT and add the result to Eq. (2.54) we obtain

$$\ln A - \ln A_R = \frac{\Delta S^\circ}{R} - \Delta n - \Delta n \ln RT \tag{2.57}$$

or

$$\frac{A}{A_R} = \left(\frac{1}{RT}\right)^{\Delta n} \exp\left(\frac{-\Delta n + \Delta S^\circ}{R}\right) \tag{2.58}$$

Here the temperature enters explicitly into the relationships, but if we again realize that kinetic data are usually available over a limited temperature range, it turns out that one can make a good correlation by taking the entropy change in the middle of the experimental temperature range to correlate average values of A and A_R (see Problem 2.14). In the few instances where accurate kinetic data are available over a wide temperature range, an equation of the form

$$k = BT^n e^{-E/RT} \tag{2.59}$$

may be used to give more flexibility than the Arrhenius equation (Problem 2.16).

References

1. See, for example, *J. Chem. Educ.* **48,** 569 (1971).
2. H. S. Johnston, *J. Amer. Chem. Soc.* **87,** 3791 (1965).
3. M. Bodenstein, *Z. Phys. Chem.* **13,** 56 (1894); **22,** 1 (1897); **29,** 295 (1899).
4. J. H. Sullivan, *J. Chem. Phys.* **30,** 1292 (1959); **46,** 73 (1967).
5. S. H. Bauer and E. Ossa, *J. Chem. Phys.* **45,** 434 (1966); K. Morokuma, L. Pedersen, and M. Karplus, *J. Amer. Chem. Soc.* **89,** 5064 (1967); L. Poulsen, *J. Chem. Phys.* **53,** 1987 (1970); R. D. Kern and G. G. Nika, *J. Phys. Chem.* **75,** 1615 (1971).
6. F. A. Lindemann, *Trans. Faraday. Soc.* **17,** 598 (1922).
7. M. Bodenstein and S. C. Lind, *Z. Phys. Chem.* **57,** 168 (1907).
8. R. Hartig, J. Troe, and H. G. Wagner, *Thirteenth Symposium (International) on Combustion,* 147 (1971).
9. A. L. Myerson and W. S. Watt, *J. Chem. Phys.* **49,** 425 (1968).
10. D. R. Stull and H. Prophet, "JANAF Thermochemical Tables," 2nd ed. U.S. Government Printing Office, Washington, D.C., 1971.
11. B. S. Rabinovitch and K. W. Michel, *J. Amer. Chem. Soc.* **81,** 5065 (1959).
12. Selected Values of Properties of Hydrocarbons and Related Compounds. American Petroleum Institute Research Project 44. Thermodynamics Research Center, Department of Chemistry, Texas A and M University, College Station, Texas.
13. E. S. Smith, M. Salkover, and H. K. Justice, "Calculus." 2d Ed. p. 215. Wiley, New York, 1958.
14. S. G. Canagaratna, *J. Chem. Educ.* **50,** 200 (1973).
15. K. J. Laidler, "Chemical Kinetics." McGraw-Hill, New York, 1965.
16. S. Arrhenius, *Z. Phys. Chem.* **4,** 226 (1889).
17. D. Graham, R. N. Haszeldine and P. J. Robinson, *J. Chem. Soc.* **B1969,** 652.
18. L. H. Gevantman and D. Garvin, *Int. J. Chem. Kinet.* **5,** 213 (1973).

Further Reading[8]

I. Amdur and G. G. Hammes, "Chemical Kinetics." McGraw-Hill, New York, 1966.

S. W. Benson, "Foundations of Chemical Kinetics." McGraw-Hill, New York, 1960.

W. C. Gardiner, "Rates and Mechanisms of Chemical Reactions." Benjamin, New York, 1969.

K. J. Laidler, "Chemical Kinetics." McGraw-Hill, New York, Second Edition, 1965.

R. E. Weston, Jr. and H. A. Schwarz, "Chemical Kinetics." Prentice-Hall, Englewood Cliffs, New Jersey, 1972.

Problems

2.1 For the elementary reaction

$$OH(g) + CO(g) \rightarrow CO_2(g) + H(g)$$

the rate constant at room temperature has been found to be approxi-

[8] Within the past few years several on-going projects involving collection and critical evaluation of rate data have been undertaken, and a number of extremely valuable reports have been issued. A descriptive survey of these projects and reports has recently been published by Gevantman and Garvin (18).

mately 7×10^7 mole^{-1} ℓ sec^{-1}. Express this rate constant in (a) mole^{-1} cc sec^{-1}; (b) mole^{-1} m^3 sec^{-1}; (c) molecule^{-1} cc sec^{-1}; (d) mole^{-1} cc min^{-1}. If a given gas mixture contains 2×10^{-10} mole ℓ^{-1} of OH radical and 3×10^{-3} mole ℓ^{-1} of CO, calculate the rate of production of CO_2 by the above reaction, in each set of units.

2.2 For the elementary reaction

$$2\,I(g) + H_2(g) \rightarrow 2\,HI(g)$$

Sullivan found a rate constant of 2.6×10^5 mole^{-2} ℓ^2 sec^{-1} at 480.7°K. If in a certain experiment the concentration of I atoms is 1.65×10^{-7} mole ℓ^{-1} and that of H_2 molecules is 2.4×10^{-3} mole ℓ^{-1}, calculate the rate of the reaction, and also $d[I]/dt$, $d[H_2]/dt$, and $d[HI]/dt$. The rate law is: rate $= k[I]^2[H_2]$.

2.3 Which of the following reactions could *possibly* be elementary (that is, occur exactly as written)?

(a) $H_2(g) + \frac{1}{2}O_2(g) \rightarrow H_2O(g)$
(b) $Zn(s) + 2HCl(aq) \rightarrow H_2(g) + ZnCl_2(aq)$
(c) $CaCO_3(s) \rightarrow CaO(s) + CO_2(g)$
(d) $Fe^{2+}(aq) + \frac{1}{2}Br_2(aq) \rightarrow Fe^{3+}(aq) + Br^-(aq)$
(e) $Fe^{3+}(aq) + Ti^{3+}(aq) \rightarrow Fe^{2+}(aq) + Ti^{4+}(aq)$
(f) $2\,C(s) + O_2(g) \rightarrow 2\,CO(g)$
(g) $C_6H_{14}(g) + \frac{19}{2}\,O_2(g) \rightarrow 6\,CO_2(g) + 7\,H_2O(g)$
(h) $Zn(H_2O)_4{}^{2+}(aq) + OH^-(aq) \rightarrow Zn(H_2O)_3(OH)^+(aq) + H_2O(aq)$

Briefly discuss the probable course of those reactions that could *possibly* be elementary, and indicate the *likelihood* of this possibility.

2.4 If a reaction is nonelementary, occurring in a series of elementary steps, then it must be true that some linear combination of the elementary reactions will yield the overall reaction (or a multiple of it) with all intermediates canceling when the equations are combined. If such a linear combination cannot be found, then the mechanism is incomplete. There is nothing wrong in principle with having more than one possible linear combination.

Several reaction mechanisms are given below. Check, by looking for linear combinations, to find which could be complete. Suggest one or more additional reactions that could occur to make the others complete.

(a) Overall Reaction

$$2\,Fe^{2+} + Tl^{3+} \rightarrow 2\,Fe^{3+} + Tl^+$$

Proposed Mechanism

$Fe^{2+} + Tl^{3+} \rightarrow Fe^{3+} + Tl^{2+}$

$Fe^{2+} + Tl^{2+} \rightarrow Fe^{3+} + Tl^{+}$

(b) Overall Reaction

$2\ NO + O_2 \rightarrow 2\ NO_2$

Proposed Mechanism

$NO + NO \rightarrow N_2O + O$

$O_2 + O \rightarrow O_3$

$O_3 + NO \rightarrow NO_2 + O_2$

(c) Overall Reaction

$2\ H_2 + O_2 \rightarrow 2\ H_2O$

Proposed Mechanism

$H_2 + O_2 \rightarrow 2\ OH$

$OH + H_2 \rightarrow H_2O + H$

$H + O_2 \rightarrow OH + O$

$O + H_2 \rightarrow OH + H$

(d) Overall Reaction

$9\ CH_3CHO \rightarrow 7\ CH_4 + 9\ CO + C_2H_6 + H_2$

Proposed Mechanism

$CH_3CHO \rightarrow CH_3 + CHO$

$CH_3 + CH_3CHO \rightarrow CH_4 + CH_3CO$

$2\ CH_3 \rightarrow C_2H_6$

$CHO \rightarrow H + CO$

$H + CH_3CHO \rightarrow CH_3CO + H_2$

2.5 In a recently prepared table of kinetic data the following rate constants were given at 298°K for two elementary reactions:

$Cl + H_2 \rightarrow HCl + H \qquad k = 8.5 \times 10^6 \ \text{mole}^{-1}\ l\ \text{sec}^{-1}$

$H + HCl \rightarrow H_2 + Cl \qquad k = 6.2 \times 10^7 \ \text{mole}^{-1}\ l\ \text{sec}^{-1}$

A table of thermodynamic data provides the following standard free energies of formation at 298°K (in units of cal mole^{-1} for a standard state of 1 atm, assuming the gases to be ideal):

Cl:	25,102	HCl:	−22,685
H_2:	0	H:	48,587

Calculate the equilibrium constant in terms of the first reaction and comment on the consistency of the two rate constants.

2.6 The dimerization of *p*-methoxybenzonitrile N-oxide ($CH_3OC_6H_4$-CNO) in carbon tetrachloride at 40°C was studied by G. Barbaro, A. Battaglia, and A. Dondoni (*J. Chem. Soc. B* **1970,** 588). For an initial concentration of 0.011 *M*, the following data were obtained:

Time (min):	0	60	120	215	325	565	942
Percent reaction:	0	9.1	16.7	26.5	32.7	47.3	60.9
Time (min):	1080	1212	1358	1518			
Percent reaction:	64.7	66.6	68.5	70.3			

Calculate the rate law and the rate constant.

2.7 O. Rogne (*J. Chem. Soc. B* **1970,** 1056) measured the rate of hydrolysis of benzenesulfonyl chloride in aqueous solution at 15°C in the presence of fluoride ion. For a fixed concentration of benzenesulfonyl chloride, initial rates of reaction were as follows:

$[F^-] \times 10^2$ (*M*)	Initial rate $\times 10^7$ (mole l^{-1} sec^{-1})
0	2.4
0.5	5.4
1.0	7.9
2.0	13.9
3.0	20.2
4.0	25.2
5.0	32.0

Note that some reaction occurs in the absence of any fluoride ion, and that this "blank" should be subtracted to give the rate produced by the reaction of interest. What is the order of the reaction with respect to $[F^-]$? Rogne reports that in this experiment the initial concentration of benzenesulfonyl chloride was 2×10^{-4} *M*, and the order with respect to it is 1. From these data write down the rate law and calculate the rate constant.

2.8 Populations of countries, animal populations, and business indicators such as the Gross National Product frequently change with time in a way that formally resembles a first-order reaction, but often with a positive exponent. For example, the population of the United States in its early years was as follows:

Year	Population
1790	3,930,000
1800	5,310,000
1810	7,240,000
1820	9,640,000
1830	12,900,000
1840	17,100,000
1850	23,200,000
1860	31,400,000

Plot a graph of log(population) versus time, and show that the population may be expressed by the equation

$$P = P_0 e^{kt}$$

where P_0 is the population in 1790. Evaluate k, and also the number of years for the population to double (this should have the same relationship to k as the half-life of a first-order reaction does). Extrapolate the population to 1970, and compare with the actual population at that time.

2.9 P. Bruck, A. Ledwith, and A. C. White (*J. Chem. Soc. B* **1970,** 205) measured the rate of reaction between triphenylmethyl hexachloroantimonate (Ph_3) and bis-(9-ethyl-3-carbazolyl) methane (**IIa**) in 1,2-dichloroethane at 40°C, obtaining the following data

$10^5 \times$ initial [Ph_3] (mole l^{-1})	$10^5 \times$ initial [**IIa**] (mole l^{-1})	Initial rate $\times 10^9$ (mole l^{-1} sec^{-1})
1.65	10.6	1.50
14.9	10.6	17.7
14.9	7.10	11.2
14.9	3.52	6.30
14.9	1.76	3.10
4.97	10.6	4.52
2.48	10.6	2.70

From these initial rate data calculate the order with respect to each of the two reactants, the overall order, and the rate constant. Compare your results with the authors' conclusion that the reaction is second order overall with an average rate constant of 1.07 $mole^{-1}$ ℓ sec^{-1}.

2.10 Calculate an equation for the half-life of a second-order reaction

$$2A \rightarrow \text{products}$$

for which rate $= k[A]^2$ in terms of the rate constant and the initial [A]. Explain why the term half-life is used mainly with first-order reactions.

2.11 The equations for first-order (2.12) and second-order (2.24) reactions look quite different, but when data are plotted the differences are not so striking.
(a) Consider a generalized reaction

$$A \rightarrow B$$

Suppose that it is first order in [A], and plot a graph of [A] versus time, analogous to Fig. 2.1, taking the initial [A] $= 1.0$ *M*, and $k = 10^{-3}$ sec. Carry the curve to 80% reaction of A. (b) Suppose now the reaction is second order. Calculate the value of a second-order rate constant that will give the same half-life for the reaction as the first-order calculation, with 1 *M* as the initial [A]. Plot [A] versus time on the same graph as the curve for (a). (c) Take the ratios of [A] for the upper and lower curve at 10, 20, . . ., 80% reaction, and obtain the average ratio. This will be an estimate of the maximum experimental error that will permit two rate laws to be distinguished when their orders differ by 1.

2.12 In this problem we will compare the results of carrying out a reaction at constant volume and constant pressure. Suppose we start with 1 ℓ of ethane at 1 atm pressure and a temperature of 800°K, at which the first-order rate for the reaction

$$C_2H_6 \rightarrow C_2H_4 + H_2$$

is about 10^{-5} sec^{-1}. Calculate the concentration of ethane as a function of time for both constant volume and constant-pressure conditions. Calculate the volume as a function of time for the constant-pressure case, and show that the number of moles or ethane decomposed at some time is equal in the two cases. (Note that this last equality would not hold in general for all reaction orders.)

2.13 Sullivan (*J. Chem. Phys.* **46,** 73 (1967)) obtained the following rate constants for the reaction

$2\,I(g) + H_2(g) \rightarrow 2\,HI(g)$

T (°K)	$10^{-5}k$ (mole^{-2} l^2 sec^{-1})
417.9	1.12
480.7	2.60
520.1	3.96
633.2	9.38
666.8	11.50
710.3	16.10
737.9	18.54

Calculate an Arrhenius equation from these data.

2.14 The rate of evolution of HCl from four samples of polyvinyl chloride was measured by V. P. Gupta and L. S. St. Pierre (*J. Polym. Sci., Part A-1* **8,** 37 (1970)) at several temperatures. The following data were obtained

Sample	Temperature (°C)	$-d(HCl)/dt \times 10^8$, mole g^{-1} sec^{-1} measured at 10% decomposition
A	215	130
	224	198
	233	339
B	215	31
	224	58
	233	336
	243	772
C	188.5	95
	206	458
	224	917
D	188.5	215
	206	263
	215	393
	224	468

Calculate an activation energy for HCl loss for each sample.

2.15 As we will show in the next chapter, the rate constant for the reaction

$H_2(g) + Ar(g) \rightarrow 2\,H(g) + Ar(g)$

may be expressed in the Arrhenius form as

$$k = 2.02 \times 10^{11}\, e^{-95{,}600/RT} \quad \text{mole}^{-1}\, \ell\, \text{sec}^{-1}$$

over the temperature range of about 2300–3800°K. By using data in the JANAF tables (or other accessible data) make the following calculations:

(a) Obtain an expression for K_c, the equilibrium constant for the reaction in units of mole ℓ^{-1}, in the form of an equation

$$K_c = C\, e^{-D/RT}$$

that is applicable over the range 2300–3800°K.

(b) Obtain an Arrhenius equation for the rate constant of the reverse reaction.

(c) Using the relationships of Section 2.10, and recognizing that for this reaction $\Delta n = 1$, calculate $\Delta S°$ and $\Delta H°$ for the reaction at the average temperature of 3000°K from the A and E values.

(d) Calculate $\Delta H°$ and $\Delta S°$ for the reaction at 3000°K directly from the JANAF tables. Compare these values with those obtained in part (c). Does the agreement show that the kinetic data are accurate? Explain your answer.

2.16 Use the approach of Section 2.10, with K_c given as in Eq. (2.52), to show that if one expresses forward and reverse rate constants in the form

$$k = BT^n\, e^{-E/RT}$$

then one can identify

$$E - E_{\mathrm{R}} = \Delta H° \qquad n_{\mathrm{R}} - n = -\Delta n \qquad \frac{B}{B_{\mathrm{R}}} = \left(\frac{1}{R}\right)^{\Delta n} \exp\left(\frac{\Delta S°}{R}\right)$$

3 Prediction of Reaction Rates

To calculate the rate constant for a reaction, one is faced with determining both A and E of an Arrhenius expression

$$k = A\,e^{-E/RT}$$

or else developing some other expression that will adequately give the rate constant over a range of temperatures. The two theoretical approaches that we will discuss—the Collision Theory and the Activated Complex Theory—concentrate on the evaluation of the Arrhenius A. At present, as we mentioned before, no theoretical approach to calculating the Arrhenius E has had extensive success, but there are some empirical rules that are useful, for lack of anything better.

3.1 COLLISION THEORY OF REACTION RATES

The kinetic theory of gases provides an approach to the evaluation of A for gaseous reactions. The theory can be applied with varying degrees of rigor, but we will use a fairly simple approach since the major limitations of the theory are not removed simply by refining it.

It will be assumed that molecules acquire the energy needed to react by the collisions they undergo. The kinetic theory can yield the number of collisions occurring in a sample of gas per unit time. Only occasionally will the kinetic energy possessed by a pair of molecules as they collide be sufficient for reaction—again, the theory can yield the fraction of collisions involving an energy greater than a certain amount. If it is assumed that all collisions with energy greater than that amount are reactive, then we can write: rate of reaction ≈ number of collisions × fraction of collisions of energy greater than E.

Let us first look at the number of collisions. If we are considering collisions of two types of molecules (as we often will for gaseous reactions), then the number of collisions that involve one molecule of each type in 1 cc per second is

$$Z_{12} = n_1 n_2 \sigma_{12}^2 \left(\frac{8\pi \mathbf{k} T}{\mu}\right)^{1/2} \tag{3.1}$$

where n_1 and n_2 are the numbers of molecules of types 1 and 2 per cubic centimeter; σ_{12} is the mean molecular diameter $(d_1 + d_2)/2$ in centimeters; $\mathbf{k}$ is the Boltzmann constant; and μ is the reduced mass of the molecules $m_1 m_2/(m_1 + m_2)$ in grams. If we wish to convert this quantity into typical kinetic units, such as mole cc^{-1} sec^{-1} rather than collisions cc sec^{-1}, we must divide Z_{12} by N, the Avogadro number. Also, we should set the number of molecules/cc equal to N times the number of moles/cc, which we can denote as $[A_1]$ and $[A_2]$. Therefore, if every collision resulted in reaction, the rate would be

$$\frac{Z_{12}}{N} = \frac{N[A_1] \cdot N[A_2] \sigma_{12}^2}{N} \left(\frac{8\pi \mathbf{k} T}{\mu}\right)^{1/2} \tag{3.2}$$

$$= N \sigma_{12}^2 \left(\frac{8\pi \mathbf{k} t}{\mu}\right)^{1/2} [A_1][A_2] \tag{3.3}$$

Let us pause to see what sort of number this gives. Values of molecular diameters may be estimated by kinetic theory methods, from gas viscosity and other properties. A typical value for a small molecule would be 4 Å, or 4×10^{-8} cm. If we consider a collision between CH_4 and O_2, for example,

$$\mu = \frac{16 \times 32}{6 \times 10^{23}(16 + 32)} = 1.78 \times 10^{-23} \quad \text{g}$$

At 1000°K,

$$N \sigma_{12}^2 \left(\frac{8\pi \mathbf{k} T}{\mu}\right)^{1/2} = 6 \times 10^{23} \times (4 \times 10^{-8})^2$$

$$\cdot \left(\frac{8 \times 3.14 \times 1.38 \times 10^{-16} \times 1000}{1.78 \times 10^{-23}}\right)^{1/2}$$

$$= 4.2 \times 10^{14} \quad \text{mole}^{-1}\ \text{cc sec}^{-1}$$

or

$$4.2 \times 10^{11} \quad \text{mole}^{-1}\ \ell\ \text{sec}^{-1}$$

These numbers indicate limiting values of rate constants; these will be approached when very little energy is needed to bring about reaction.

When we turn to the energetics of the collisions, it is clear that not only the relative velocities, but also the detailed trajectories of molecular pairs will differ. One would expect that, for a given speed of approach, reaction would be more likely in the case of a head-on collision than a glancing one. It is convenient, and not greatly inaccurate, to calculate only the fraction of head-on collisions that leads to reaction. For these, each molecule is moving directly toward their center of mass, and if the velocities with respect to the center of mass are v_1 and v_2, then conservation of momentum requires that

$$m_1v_1 + m_2v_2 = 0 \tag{3.4}$$

while the kinetic energy is

$$KE = \tfrac{1}{2}m_1v_1^2 + \tfrac{1}{2}m_2v_2^2 \tag{3.5}$$

These two equations can be combined to give a new equation for the kinetic energy

$$KE = \tfrac{1}{2}\mu c^2 \tag{3.6}$$

where μ is the reduced mass of the molecules and c is $v_1 - v_2$ (where v_1 and v_2 have opposite signs so that c is the speed of approach of the molecules).

We are really considering a system with only two degrees of translational motion (the motion of each molecule along the line joining the centers) and the kinetic theory provides an equation for the fraction of times the speed lies between c and $c + dc$

$$\frac{dn}{n} = \left(\frac{\mu}{\mathbf{k}T}\right)\exp\left(\frac{-\mu c^2}{2\mathbf{k}T}\right)c\,dc \tag{3.7}$$

If we wish to use this equation in energy terms, and particularly in energy per mole, we note that

$$E = \tfrac{1}{2}N\mu c^2, \qquad dE = N\mu c\,dc, \qquad \text{and} \qquad \mathbf{k} = R/N$$

so that

$$\frac{dn}{n} = \frac{1}{RT}e^{-E/RT}\,dE \tag{3.8}$$

The fraction of collisions with energy above a certain amount, then, is

$$\int_{E_{\mathrm{CT}}}^{\infty} \frac{1}{RT}e^{-E/RT}\,de = \frac{1}{RT}(RT)e^{-E/RT}\Big|_{E_{\mathrm{CT}}}^{\infty} = \exp(-E_{\mathrm{CT}}/RT) \tag{3.9}$$

Thus the rate constant according to the collision theory (with the approximations we made) is

$$k = N\sigma_{12}^2\left(\frac{8\pi\mathbf{k}T}{\mu}\right)^{1/2} \exp\left(\frac{-E_{\mathrm{CT}}}{RT}\right) \tag{3.10}$$

Since there is a $T^{1/2}$ term in the collision number, we must make a rearrangement like that necessary for thermodynamic considerations in order to get an Arrhenius equation, and we will expect the Arrhenius E to be a little different from E_{CT}. Write

$$k = BT^{1/2} \exp\left(\frac{-E_{\mathrm{CT}}}{RT}\right) = Ae^{-E/RT}$$

where

$$B = N\sigma_{12}^2\left(\frac{8\pi\mathbf{k}}{\mu}\right)^{1/2}$$

Take logarithms

$$\ln B + \frac{1}{2}\ln T - \frac{E_{\mathrm{CT}}}{RT} = \ln A - \frac{E}{RT}$$

and differentiate with respect to T

$$\frac{1}{2T} + \frac{E_{\mathrm{CT}}}{RT^2} = \frac{E}{RT^2}$$

$$\boxed{E = E_{\mathrm{CT}} + \tfrac{1}{2}RT} \tag{3.11}$$

$$\ln B + \tfrac{1}{2}\ln T + \tfrac{1}{2} = \ln A$$

$$\boxed{A = e^{1/2}BT^{1/2}} \tag{3.12}$$

These equations will apply to reactions involving two molecules of the same substance reacting together, except that if the stoichiometric coefficient is two,

$$k = N\sigma^2\left(\frac{2\pi\mathbf{k}T}{m}\right)^{1/2} \exp\left(\frac{-E_{\mathrm{CT}}}{RT}\right) \tag{3.13}$$

since the rate of reaction is half the rate of disappearance of this substance. The equations will also apply to reactions in which the rate of reactions depends on a collision between two molecules, even if only one reacts. For example, Myerson and Watt (1) expressed their results for the reaction

$$H_2(g) + Ar(g) \rightarrow 2\,H(g) + Ar(g)$$

in collision theory form as

$$k = 2.23 \times 10^{12} T^{1/2} e^{-92,600/RT} \quad \text{cm}^3\ \text{mole}^{-1}\ \text{sec}^{-1}$$

the data extending over the range 2290–3790°K. If we wish to express these results in Arrhenius form, we could take an average temperature of 3000°K and write

$$E = 92,600 + \tfrac{1}{2} \times (2 \times 3000) = 95,600$$
$$A = 1.65 \times 2.23 \times 10^{12} \times 0.548 \times 10^2 = 2.02 \times 10^{14}$$

or

$$k = 2.02 \times 10^{14} e^{-95,600/RT} \quad \text{mole}^{-1}\ \text{cc sec}^{-1}$$
$$k = 2.02 \times 10^{11} e^{-95,600/RT} \quad \text{mole}^{-1}\ \ell\ \text{sec}^{-1}$$

The Arrhenius equation and the collision theory equation give rate constants differing by no more than 2% over the experimental temperature range, so are experimentally indistinguishable since the experimental scatter in the data is above 10%. Let us see how closely the experimental B value agrees with theory.

From gas viscosity, molecular diameters of 2.9 Å for argon and 2.2 Å for H_2 have been found, so $\sigma_{12} = 2.55 \times 10^{-8}$ cm. The reduced mass μ is given by

$$\mu = \frac{2 \times 40}{6 \times 10^{23}(2 + 40)} = 3.17 \times 10^{-24} \quad \text{g}$$

Accordingly,

$$B = N\sigma_{12}^2\left(\frac{8\pi\mathbf{k}}{\mu}\right)^{1/2} = 6 \times 10^{23} \times (2.55 \times 10^{-8})^2$$

$$\cdot\left(\frac{8 \times 3.14 \times 1.38 \times 10^{-16}}{3.17 \times 10^{-24}}\right)^{1/2}$$

$$= 1.29 \times 10^{13} \quad \text{cm}^3\ \text{mole}^{-1}\ \text{sec}^{-1}\ \text{deg}^{-1/2}$$

The difference of a factor of 5 between the observed and calculated values of B is not unusual for this type of comparison. Indeed, it seems reasonable that not all sufficiently energetic collisions between Ar and H_2 molecules would lead to reaction. Probably a collision in which the Ar struck the H_2 broadside would be less effective in exciting vigorous H—H motion than one in which the Ar struck the molecule end-on.

A factor p was soon introduced into the collision theory expression, to give

$$k = pBT^{1/2}\exp\left(\frac{-E_{CT}}{RT}\right) \tag{3.14}$$

where p is the probability factor, assumed to be somewhat less than 1. Kinetic data on a number of bimolecular gas reactions are given in Table 3.1. The collision theory is in order-of-magnitude agreement with the Arrhenius A values for most of these reactions, and also gives the correct trends. For example, reactions involving the fast-moving H atom tend to have high A values. A values for dissociation reactions of polyatomic molecules such as HO_2 tend to be high because internal vibrational degrees of freedom of the dissociating molecule can contribute energy to the dissociation process. On the other hand, when two complicated molecules react with each other, the A value may be quite low. The effect even shows up somewhat when two diatomic molecules react (see $CO + OH$ and $NO + O_2$ in Table 3.1) and tends to increase with the complexity of the reaction process. In other words, the simple collision theory becomes less accurate as processes other than the intermolecular collision itself become important.

Table 3.1 Kinetic Data for Several Bimolecular Gas Reactions[a]

Reactions	A (mole^{-1} l sec^{-1})	E (kcal)[b]	ΔH (kcal)[b]
$O + H_2 \rightarrow OH + H$	1.7×10^{10}	9	2
$OH + H_2 \rightarrow H_2O + H$	2×10^{10}	5	−15
$CO + OH \rightarrow CO_2 + H$	6×10^{8}	1	−23
$H + CO_2 \rightarrow CO + OH$	6×10^{10}	24	23
$H + H_2O \rightarrow H_2 + OH$	8×10^{10}	20	15
$O + H_2O \rightarrow 2\,OH$	6×10^{10}	18	17
$OH + OH \rightarrow H_2O + O$	6×10^{9}	1	−17
$H + O_2 \rightarrow OH + O$	2×10^{11}	17	17
$O + OH \rightarrow O_2 + H$	1.3×10^{10}	0	−17
$HO_2 + Ar \rightarrow H + O_2 + Ar$	2×10^{12}	46	47
$N + O_2 \rightarrow NO + O$	6×10^{9}	6	−32
$NO + N \rightarrow N_2 + O$	3×10^{10}	0	−75
$O + N_2 \rightarrow NO + N$	1.4×10^{11}	75	75
$O + NO \rightarrow N + O_2$	1.6×10^{9}	38	32
$NO + O_2 \rightarrow NO_2 + O$	1×10^{9}	46	46

[a] Data from "High Temperature Reaction Rate Data," Nos. 1–5, D. L. Baulch, D. D. Drysdale, and A. C. Lloyd. The University, Leeds 2, England, 1968–1970.

[b] Values are averages over the temperature range, which is typically 300–2000°K.

Recently, the collision theory has been reformulated in terms analogous to those used in nuclear chemistry, with a "reaction cross section" in place of the value of σ^2 from the kinetic theory (2). The reaction cross section is taken to be a parameter that may vary considerably with the energy of the collision and need not be even of the same order of magnitude as σ^2. This approach permits the collision theory to "explain" a wider range of kinetic data, and it is surely right in principle (as we will see in some detail later) that the probability of reaction is energy dependent. However, the collision theory itself does not give any indication of what values the reaction cross sections should have under given circumstances, so this formulation is not, as yet, useful in predicting reaction rates.

3.2 INTRODUCTION TO ACTIVATED COMPLEX THEORY

The Activated Complex Theory of reaction rates was put forward by Henry Eyring in 1935 (3), being called at that time the Absolute Reaction Rate Theory. This name is somewhat misleading since, like the Collision Theory, Eyring's theory is not able to predict the activation energy. For this reason, and because the theory focuses attention on the nature of the activated complex, we prefer the name Activated Complex Theory, or ACT. None of the derivations of the theory (of which there are several) is free from approximations, but the fact that the theory has accounted for a great deal of experimental fact has given it an empirical validity within recognized limitations. It is at present the most useful of reaction rate theories, being applicable to condensed-phase as well as gaseous reactions.

In simplest terms, the ACT assumes that an equilibrium exists between reactants and the activated complex, and it provides a method for calculating the product of the equilibrium constant and the rate constant for the conversion of activated complex into product. Since the activated complex is a transitory species present in small amount, the equilibrium constant cannot be measured experimentally, so the method used is one of statistical thermodynamics, in which partition functions of the reactants and the activated complex are combined with an estimate of the zero-point energy difference between complex and reactants.

Because an understanding of this thermodynamic method and a feel for the molecular parameters that enter into its use are essential in order to use the theory, these two topics are discussed in the next two sections. A student who is already familiar with this material should feel free to skim rapidly through these sections, perhaps test his understanding by trying Problem 3.7, and then go on to Section 3.5.

3.3 PARTITION FUNCTIONS AND THE EQUILIBRIUM CONSTANT

The partition function arises from the quantum theory concept that a molecule or chemical system can exist only in states with certain definite energies. The partition function is sometimes called the "sum of states." Perhaps it can best be introduced by way of an example.

Suppose that a container holds molecules of substance A and substance B that are in equilibrium with one another. Under these conditions we can write an equilibrium constant in terms of concentrations, K_c, as

$$K_c = \frac{[\mathrm{B}]}{[\mathrm{A}]}$$

for the reaction A $\rightleftarrows$ B. Since in this example both substances are contained in the same volume, we can write

$$K_c = \frac{N_{\mathrm{B}}}{N_{\mathrm{A}}}$$

where N_{B} and N_{A} are the numbers of molecules of each type.

An energy diagram for this system is shown schematically in Fig. 3.1. Molecules of type A have, of course, a lowest energy level, which is denoted by $E_{0\mathrm{A}}$. Its energy is taken as 0. Molecules of type B also have a lowest energy level $E_{0\mathrm{B}}$, which is also taken as 0 as far as molecules of type B are concerned. However, $E_{0\mathrm{B}}$ is higher than $E_{0\mathrm{A}}$ by an amount $E_{0\mathrm{R}}$. $E_{0\mathrm{R}}$, therefore, is the energy needed to carry a molecule from the lowest energy level of A molecules to the lowest energy of B molecules, or, in other words, the zero-point energy difference. This latter term arises from the idea that, at 0°K, all of the molecules would be in their lowest energy states.

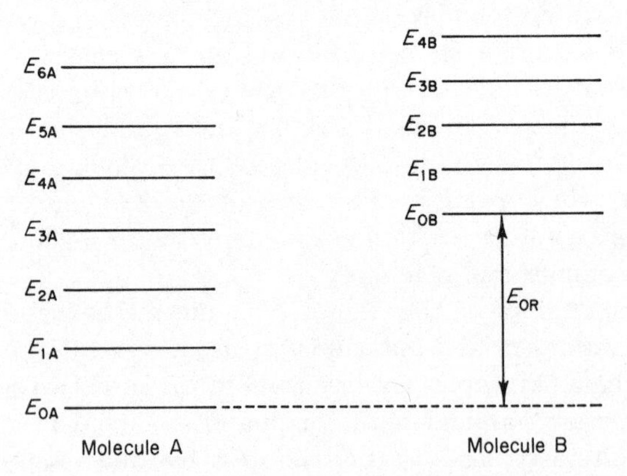

Fig. 3.1 Schematic energy level diagram for the reaction A $\rightleftarrows$ B.

Each energy level may correspond to more than one quantum state. The number of quantum states that give rise to a particular energy level, that is, the degeneracy or multiplicity of the energy level, is denoted by g. A subscript denotes the energy level, so that $g_{3\mathrm{B}}$ is the multiplicity of level 3 for B molecules.

For any two energy levels, the ratio of the number of molecules in each is given by the Boltzmann distribution

$$\frac{N_2}{N_1} = \frac{g_2 \exp(-E_2/\mathbf{k}T)}{g_1 \exp(-E_1/\mathbf{k}T)} \tag{3.15}$$

where $\mathbf{k}$ is the Boltzmann constant, T the absolute temperature, and E_1 and E_2 are in ergs per molecule if the Boltzmann constant is in its traditional units of erg deg K^{-1}. Therefore,

$$N_2 = N_1 \frac{g_2}{g_1} \exp\left(\frac{-(E_2 - E_1)}{\mathbf{k}T}\right) \tag{3.16}$$

We are now in a position to write expressions for the numbers of molecules in each energy state. It will be convenient to do so in terms of $N_{0\mathrm{A}}$, the number of molecules in the level with $E_{0\mathrm{A}}$. Let us start with the molecules of type A:

$$N_{\mathrm{A}} = N_{0\mathrm{A}} + N_{0\mathrm{A}} \frac{g_{1\mathrm{A}}}{g_{0\mathrm{A}}} \exp\left(\frac{-(E_{1\mathrm{A}} - 0)}{\mathbf{k}T}\right) + N_{0\mathrm{A}} \frac{g_{2\mathrm{A}}}{g_{0\mathrm{A}}} \exp\left(\frac{-(E_{2\mathrm{A}} - 0)}{\mathbf{k}T}\right) + \cdots \tag{3.17}$$

Multiply by $g_{0\mathrm{A}}$ and factor out $N_{0\mathrm{A}}$

$$g_{0\mathrm{A}} N_{\mathrm{A}} = N_{0\mathrm{A}} \left[g_{0\mathrm{A}} + g_{1\mathrm{A}} \exp\left(\frac{-E_{1\mathrm{A}}}{\mathbf{k}T}\right) + g_{2\mathrm{A}} \exp\left(\frac{-E_{2\mathrm{A}}}{\mathbf{k}T}\right) + \cdots \right] \tag{3.18}$$

Note that since $E_{0\mathrm{A}} = 0$, we could write

$$g_{0\mathrm{A}} = g_{0\mathrm{A}} \exp\left(\frac{-E_{0\mathrm{A}}}{\mathbf{k}T}\right) \tag{3.19}$$

to get it into the same formal representation as the other terms. Thus,

$$g_{0\mathrm{A}} N_{\mathrm{A}} = N_{0\mathrm{A}} \sum_{i=0}^{\infty} g_{i\mathrm{A}} \exp\left(\frac{-E_{i\mathrm{A}}}{\mathbf{k}T}\right) \tag{3.20}$$

where the summation is carried over all energy levels. At any temperature, the value of exp $(-E_{iA}/\mathbf{k}T)$ will diminish as we go to higher energies, and eventually the terms will become negligibly small as far as our desired accuracy is concerned. Finally then,

$$N_A = \frac{N_{0A}}{g_{0A}} \sum_i g_{iA} \exp\left(\frac{-E_{iA}}{\mathbf{k}T}\right) \tag{3.21}$$

The quantity

$$\sum_i g_{iA} \exp\left(\frac{-E_{iA}}{\mathbf{k}T}\right) \tag{3.22}$$

is the partition function for substance A, and is usually denoted by Q or F.

In a similar way, we can calculate the number of molecules of B in its various energy levels. For comparison with N_A, it should be calculated in terms of N_{0A}. The energies of the various states for molecule B, with E_{0A} as the starting point, are (see Fig. 3.1) $E_{0R} + E_{0B}$ (where $E_{0B} = 0$), $E_{0R} + E_{1B}$ and so on. Thus,

$$N_B = N_{0A} \frac{g_{0B}}{g_{0A}} \exp\left(\frac{-(E_{0R} + E_{0B})}{\mathbf{k}T}\right) + N_{0A} \frac{g_{1B}}{g_{0A}} \exp\left(\frac{-(E_{0R} + E_{1B})}{\mathbf{k}T}\right) + \cdots \tag{3.23}$$

$$= \frac{N_{0A}}{g_{0A}} \exp\left(\frac{-E_{0R}}{\mathbf{k}T}\right)\left(g_{0B} \exp\left(\frac{-E_{0B}}{\mathbf{k}T}\right) + g_{1B} \exp\left(\frac{-E_{1B}}{\mathbf{k}T}\right) + \cdots\right) \tag{3.24}$$

$$= \frac{N_{0A}}{g_{0A}} \exp\left(\frac{-E_{0R}}{\mathbf{k}T}\right) \sum_i g_{iB} \exp\left(\frac{-E_{1B}}{\mathbf{k}T}\right) \tag{3.25}$$

Therefore

$$K = \frac{N_B}{N_A} = \frac{(N_{0A}/g_{0A}) \exp(-E_{0R}/\mathbf{k}T) \sum_i g_{iB} \exp(-E_{iB}/\mathbf{k}T)}{(N_{0A}/g_{0A}) \sum_i g_{iA} \exp(-E_{iA}/\mathbf{k}T)} \tag{3.26}$$

$$= \frac{Q_B}{Q_A} \exp\left(\frac{-E_{0R}}{\mathbf{k}T}\right) \tag{3.27}$$

If we change E_{0R} (ergs per molecule) to ΔE_0 (cal per mole) and $\mathbf{k}$ (erg deg^{-1}) to R (cal deg^{-1}), then we get

$$K = \frac{Q_B}{Q_A} \exp\left(\frac{-\Delta E_0}{RT}\right) \quad \text{or} \quad K = \frac{F_B}{F_A} \exp\left(\frac{-\Delta E_0}{RT}\right) \tag{3.28}$$

What we have accomplished is the expression of the equilibrium constant in terms of two quantities Q_A and Q_B that are properties of the individual reactants, and one quantity ΔE_0 that is characteristic of the reaction.

The equilibrium constant is, of course, closely related to the free energy, and we may write

$$\begin{aligned} \Delta G^\circ &= (G_B{}^\circ - E_B{}^\circ) - (G_A{}^\circ - E_A{}^\circ) + \Delta E_0 \\ &= -RT \ln K \\ &= (-RT \ln Q_B) - (-RT \ln Q_A) + \Delta E_0 \end{aligned} \tag{3.29}$$

Equation (3.29) suggests that the free energy may be equal to $-RT \ln Q$. This is true for energy levels such as those we have been considering (where each molecule has a set that it can occupy independently of the others) but is not correct for translational energy levels of a gas (where all the molecules rather thinly populate a single set of many energy levels). In the next few pages we will look at the various types of energy levels molecules can have, and the relationships of the partition functions to the free energy.

Before doing this, however, we must consider how to combine sets of energy levels to obtain the total partition function. For example, suppose a molecule has one set of rotational energy levels and one set of vibrational energy levels, which can be occupied independently of one another. We could make a complete list of energy levels consisting of all possible combinations of rotational and vibrational energy levels, then add up the contributions of all to get the partition function, but there is an easier way. Since a molecule in each given vibrational state may occupy any rotational state, the sum of all the states (allowing for the Boltzmann energy effect) will be $Q_V Q_R$, where Q_V and Q_R are the partition functions for the vibrational and rotational levels as separate groups. This principle will also apply when two molecules are reacting, and we are interested in the total partition function for the pair. If the molecules are A and B, then this total partition function will be $Q_A Q_B$ since each molecule can occupy its energy levels independently of the other.

Real molecules may have translational, rotational, vibrational, and electronic energies, and sometimes other kinds (such as internal rotations). The effect of interaction of one type with another on the partition function is usually small (often less than 10% and seldom more than a factor of 2) and the simplifications produced by neglecting interactions are great. Since we will have to do some estimating with regard to the energy levels in the activated complex anyway, we will make the simplification, and write

$$Q = Q_T Q_R Q_V Q_E \tag{3.30}$$

where the subscripts indicate the type of energy level.

3.4 CALCULATION OF PARTITION FUNCTIONS

3.1.4 Translational Partition Functions

For the translational motion of gases, the Schrödinger equation for the energy levels of a particle in a box gives a point of departure. For motion in one dimension,

$$E = \frac{n^2 h^2}{8ma^2} \tag{3.31}$$

where n is a quantum number $(1, 2, 3, 4, \ldots)$, h is Planck's constant, m the mass of the particle, and a the length of the box. We can calculate Q for one dimension, then cube it to get the total translational partition function. For this one dimension,

$$Q = \sum_i \exp\left(\frac{-n_i^2 h^2}{8ma^2 \mathbf{k} T}\right) \tag{3.32}$$

since the multiplicities of the one-dimensional energy levels are all equal to 1. In considering a macroscopic quantity of gas, we recognize that these energy levels are very closely spaced[1] and that the sum may therefore be replaced by an integration

$$Q = \int_0^\infty \exp\left(\frac{-n_i^2 h^2}{8ma^2 \mathbf{k} T}\right) dn_i = \frac{(2\pi m \mathbf{k} T)^{1/2} a}{h} \tag{3.33}$$

The total translational partition function in three dimensions is then

$$Q_T = \frac{(2\pi m \mathbf{k} T)^{3/2} a^3}{h^3} \tag{3.34}$$

We may recognize that $a^3 = V$, the volume of the gas. If our standard state is 1 mole cc^{-1}, then $a^3 = 1$ cc, while if the standard state is 1 mole ℓ^{-1},

[1] You may check this point by calculating the energy levels for a few values of n, using the mass in grams of a typical molecule and a length a of $\sim$10 cm. Compare these energies with the average translational energy of a molecule in one dimension, $\frac{1}{2}\mathbf{k}T$.

$a^3 = 1000$ cc (not 1 ℓ, since the units of a^3 must match those of the other physical constants to make Q_T dimensionless).

In terms of free energy and the equilibrium constant, the quantity we want to use is not Q_T itself, but Q_T/N, so that

$$(G - E_0°)_T = -RT \ln \frac{Q_T}{N} \tag{3.35}$$

where N is the Avogadro number. This comes about because in calculating the free energy and the equilibrium constant we are really interested in the number of ways the molecules can be arranged among the states, not the sum of states per se. When we distribute N molecules among a set of states, we get many arrangements differing only in that different molecules occupy given levels, and are therefore indistinguishable. When the number of available energies is considerably larger than the number of molecules (as is true unless the pressure becomes quite high), the number of distinguishable arrangements becomes Q_T/N. Since for practical purposes we will use this ratio rather than Q_T itself, let us denote it by Q_T'. If the numerical values of the constants are substituted into Eq. (3.34), and we use the molecular weight M in place of molecular mass m ($M = mN$), then we obtain a working equation

$$Q_T' = \frac{Q_T}{N} = 0.000311 T^{3/2} M^{3/2} V \tag{3.36}$$

where V, as indicated above, is either 1 cc or 1000 cc. For example, for a standard state of 1 mole ℓ^{-1} at room temperature, Q_T' for N_2 gas is

$$Q_T' = 0.000311(298)^{3/2}(28)^{3/2}(1000) = 2.3 \times 10^5$$

We may note that there will be no difficulty in obtaining Q_T' for a gaseous activated complex since the only molecular parameter is the molecular weight. Also, for any unimolecular reaction in which the activated complex is a specially energized molecule, the translational contributions to the total partition function will cancel out of the equilibrium constant expression.

Translational partition functions for liquids are more difficult to calculate since there will probably be considerable interaction among the molecules, and molecules are certainly not as free to move around as they are in the gas phase. One approach is to suppose that each molecule is in a small "box" bounded by neighboring molecules, and moves independently of the other molecules within this box. In this situation

$$(G° - E_0°)_T = -RT \ln Q_T \tag{3.37}$$

where $Q_T = 1.8 \times 10^{20} T^{3/2} M^{3/2} V$ and V is now of molecular dimensions. For example, suppose we consider that acetone molecules occupy a "box" for which the volume can be calculated from the density and molecular weight:

$$V = \frac{M}{\rho N} = \frac{58}{0.79 \times 6 \times 10^{23}} = 1.22 \times 10^{-22} \quad \text{cc}$$

Accordingly, at room temperature

$$Q_T = 1.8 \times 10^{20}(298)^{3/2}(58)^{3/2}(1.22 \times 10^{-22}) = 5 \times 10^4$$

This calculation is clearly a very approximate one. A more common approach for liquids is to estimate Q_T from thermodynamic data on either the compound of interest or some similar substance.

3.4.2 Rotational Partition Functions

For a linear rigid rotor, the energy levels are given by the expression

$$E = J(J + 1)B \tag{3.39}$$

where J is a quantum number ($J = 0, 1, 2, \ldots$) and B a constant depending on the structure of the molecule. The multiplicity of each energy level is $2J + 1$, so

$$Q_R = \sum_J (2J + 1)e^{-J(J+1)B/\mathbf{k}T} \approx \int_0^\infty (2J + 1)e^{-J(J+1)B/\mathbf{k}T}\, dJ \tag{3.40}$$

where the integral will be a good approximation if the energy levels are small compared to $\mathbf{k}T$, a valid one for most molecules above room temperature.

The form of B depends on the geometry of the molecule. For linear molecules that are not symmetrical (such as CO, HCl, NNO, HCN)

$$B = \frac{h^2}{8\pi^2 I} \tag{3.41}$$

where I is the moment of inertia. When this expression is put into Eq. (3.40), one obtains

$$Q_R = \frac{8\pi^2 I \mathbf{k}T}{h^2} \tag{3.42}$$

The moment of inertia for any molecule (about an axis passing through the center of gravity) is given by the general formula

$$I = \sum_i m_i r_i^2 \tag{3.43}$$

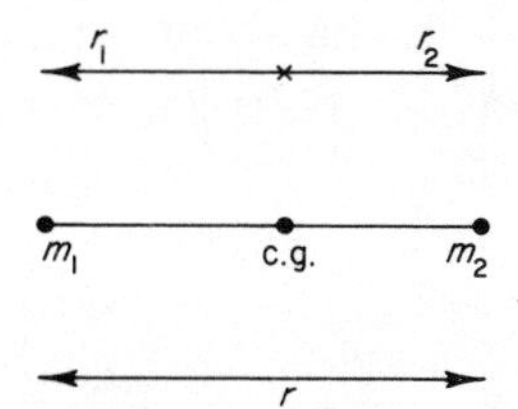

Fig. 3.2 Calculation of the moment of inertia of a diatomic molecule.

where m_i is the mass of an atom and r_i its distance from the axis. For the special case of a linear molecule, I reduces to a relatively simple relationship (see Fig. 3.2) The center of gravity is the balance point, so $m_1r_1 = m_2r_2$, where $r_1 + r_2 = r$. It can easily be shown that

$$I = \mu r^2 \tag{3.44}$$

where μ is the reduced mass, $m_1m_2/(m_1 + m_2)$. If we use the traditional units of grams for atomic masses and centimeters for distances, moments of inertia are typically on the order of 10^{-40} to 10^{-37} g cm^2.

For symmetrical linear molecules, such as H_2, N_2, OCO, and HCCH, the actual partition functions are obtained by dividing the result of Eq. (3.42) by 2. The factor of 2 is the symmetry number, usually denoted by σ, and is the number of ways the molecule can be positioned and still look the same. The symmetry represented by σ causes certain rotational energy levels to be absent, thereby reducing the partition function, and having the appropriate effect on the equilibrium constant. For example, in the reaction

$$^{35}Cl^{35}Cl + {}^{37}Cl^{37}Cl \rightleftarrows 2\,{}^{35}Cl^{37}Cl$$

the symmetry numbers cause the equilibrium constant to be nearly 4 instead of nearly 1, and they affect rate constants also. As we will see in Section 3.5, the best way to account for this type of symmetry effect on reaction rates is by the "statistical factor" rather than by use of σ, and we will find it convenient to define

$$Q_R' = Q_R\sigma \tag{3.45}$$

and use this modified partition function in the same way that we will use Q_T'. However, in this instance the correct relationship between partition function and free energy is

$$(G^\circ - E_0^\circ)_R = -RT ln Q_R \tag{3.46}$$

and free energies tabulated in handbooks will include the symmetry number.

A nonlinear molecule can rotate around three mutually perpendicular axes, and there may be a different moment of inertia about each axis. If the three moments of inertia are A, B, and C, the partition function for a nonlinear molecule is

$$Q_R = \frac{8\pi^2(8\pi^3 ABC)^{1/2}(\mathbf{k}T)^{3/2}}{\sigma h^3} \tag{3.47}$$

If numerical values of the constants are put into Eqs. (3.42) and (3.47), we obtain

$$\text{for linear molecules:} \quad Q_R = \frac{0.0244(I \times 10^{40})T}{\sigma} \tag{3.48}$$

$$\text{for nonlinear molecules:} \quad Q = \frac{6.69 \times 10^{-3}(ABC \times 10^{120})^{1/2}T^{3/2}}{\sigma} \tag{3.49}$$

Calculation of these partition functions, then, requires some knowledge or at least an estimate of the molecular geometry, which for the complex will have to come from our knowledge of general chemistry.

Usually, molecules can rotate freely in liquids, so the same calculations would carry over to them. In solids, though, molecules may be prevented from rotating by the relatively strong intermolecular forces present.

3.4.3 Vibrational Partition Functions

Vibrational energy levels are usually so far apart that one must sum rather than integrate them. Fortunately this is not hard to do. For a simple harmonic oscillator, the vibrational energy is

$$E_V = (v + \tfrac{1}{2})E \tag{3.50}$$

where E is the energy spacing between the energy levels, and v is a quantum number 0, 1, 2, ... Each level has a multiplicity of 1, so one can write

$$Q_V = 1 + e^{-E/\mathbf{k}T} + e^{-2E/\mathbf{k}T} + \cdots \tag{3.51}$$

This sum can be obtained by multiplying both sides by $e^{-E/\mathbf{k}T}$ and subtracting

$$e^{-E/\mathbf{k}T}Q_V = e^{-E/\mathbf{k}T} + e^{-2E/\mathbf{k}T} + \cdots \tag{3.52}$$

$$Q_V - e^{-E/\mathbf{k}T}Q_V = 1 \tag{3.53}$$

since all the other terms cancel. This equation can be arranged to give

$$Q_V = \frac{1}{1 - e^{-E/\mathbf{k}T}} \tag{3.54}$$

In this equation E must be in ergs. By Planck's equation,

$$E = h\nu = hc\tilde{\nu} \tag{3.55}$$

where ν is the frequency in $\sec^{-1}$ and $\tilde{\nu}$ that in cm^{-1}, probably the most common unit of practical vibrational spectroscopy. Numerically,

$$E \quad (\text{erg}) = 1.986 \times 10^{-16}\,\tilde{\nu} \quad (\text{cm}^{-1}) \tag{3.56}$$

If the molecule has several degrees of vibrational freedom, the vibrational partition function will be the product of those for each frequency.

Let us note a limiting case we will use later. If the vibrational energy level spacing is small compared to $\mathbf{k}T$, then

$$Q_{\text{V}} \approx \frac{1}{1 - (1 - E/\mathbf{k}T)} = \frac{\mathbf{k}T}{E} = \frac{\mathbf{k}T}{h\nu} \tag{3.57}$$

3.4.4 Electronic Partition Functions

There are usually few, if any, electronic energy levels other than the ground state to consider because in most molecules electronic energy levels are widely spaced. If there are any, the partition function is normally obtained by direct summation as

$$Q_{\text{E}} = \sum_i g_i \exp\left(\frac{-E_{\text{E}}}{\mathbf{k}T}\right) \tag{3.58}$$

A common situation, which would apply in particular if there are unpaired electrons in the molecule, is that g_0 may be greater than 1. Since this effect could cause a change in the calculated rate constant of a factor of 2 or more, thought should be given to the probable multiplicity of the ground state of the activated complex.

3.4.5 A Sample Calculation

Let us calculate the partition function for the CO_2 molecule at 1500°K, at a standard state of 1 mole ℓ^{-1}.

Translational Part We can use Eq. (3.36) to get Q_{T}', the only additional information needed being the molecular weight, 44:

$$Q_{\text{T}}' = 0.000311(1500)^{3/2}(44)^{3/2}(1000) = 5.60 \times 10^6$$

Rotational Part CO_2 is a linear molecule, the C—O distances being 1.16 Å. Since the carbon atom is at the center of gravity, we can write, by

Eq. (3.43),

$$I = \sum_i m_i r_i^2 = 2m_O(1.16 \times 10^{-8})^2$$

where m_O is the mass of the oxygen atom.

$$I = \frac{2 \times 16}{6.02 \times 10^{23}} (1.16 \times 10^{-8})^2 = 71.5 \times 10^{-40} \text{ g cm}^2$$

The symmetry number is 2, so the rotational partition function, by Eq. (3.48), is

$$Q_R = \frac{0.0244 \times 71.5 \times 1500}{2} = 1.31 \times 10^3$$

Vibrational Part CO_2 has three frequencies, two involving bond stretching and compression with frequencies of 1343 and 2349 cm^{-1}, and one involving bond bending which is doubly degenerate (to allow bending in all directions) with frequency 667 cm^{-1}. We will note that the value of the Boltzmann constant is 1.38×10^{-16} erg deg^{-1}. Perhaps the calculations can best be given as in Table 3.2. The total vibrational part of the partition

Table 3.2

$\tilde{\nu}$ (cm^{-1})	E (erg)	$E/\mathbf{k}T$	$e^{-E/\mathbf{k}T}$	Q_V
	$\times 10^{13}$			
1343	2.67	1.29	.275	1.38
2349	4.67	2.26	.104	1.12
667	1.33	0.64	.527	2.11

function will be

$$Q_V = 1.38 \times 1.12 \times 2.11^2 = 6.90$$

It is interesting to note that if the temperature had been much lower (say room temperature), the value of Q_V would be approaching 1. Frequently vibrational contributions can be neglected in low-temperature calculations—a considerable savings of time being effected.

Electronic Part CO_2 has a singlet ground state, with all electrons paired, and has no low-lying excited electronic states, so that $Q_E = 1.00$.

Accordingly, the total partition function is given by

$$\frac{Q}{N} = (5.60 \times 10^6) \times (1.31 \times 10^3) \times (6.90) \times (1.00) = 5.1 \times 10^{10}$$

3.4.6 Partition Functions from Tabulated Thermodynamic Data

Although partition functions are useful, there has been no major collection of them prepared for the use of kineticists. However, partition functions for many substances, particularly for gases, may be obtained from free energies, which are tabulated in a number of places, particularly the JANAF (4) and API (5) tables. For gases, we have

$$G^\circ - E_0^\circ = -RT \ln \frac{Q}{N} \tag{3.59}$$

the standard state being the ideal state at 1 atm. Since the partition function is proportional to the volume, by Eq. (3.38), we may convert to any other standard state by multiplying by the appropriate ratio of volumes.

As an example, let us calculate the partition function of CO_2 at 1500°K and a standard state of 1 mole ℓ^{-1}, using information from the JANAF tables. The free energy at 1500°K is given in the form

$$-\frac{F^\circ - H_{298}^\circ}{T} = 59.984 \quad \text{cal deg}^{-1}\ \text{mole}^{-1}$$

where F is an old-fashioned way of expressing the Gibbs free energy G. Since we are interested in $G^\circ - E_0^\circ$ (or $G^\circ - H_0^\circ$ since at 0°K the internal energy E_0° and the enthalpy H_0° of a perfect gas become equal) we combine this information with an additional entry in the table:

$$H_0^\circ - H_{298}^\circ = -2.238 \quad \text{kcal mole}^{-1}$$

Let us write

$$G^\circ - H_{298}^\circ = -1500 \times 59.984 = -89{,}976 \quad \text{cal mole}^{-1}$$

$$H_{298}^\circ - H_0^\circ = +2{,}238 \quad \text{cal mole}^{-1}$$

Adding,

$$G^\circ - H_0^\circ = -87{,}738 \quad \text{cal mole}^{-1}$$

By Eq. (3.59)

$$\ln \frac{Q}{N} = \frac{87{,}738}{1.987 \times 1500} = 29.437$$

$$\log \frac{Q}{N} = 12.782$$

$$\frac{Q}{N} = 6.06 \times 10^{12}$$

This value of the partition function applies at 1 atm, and at 1500°K the volume of a mole of gas at this pressure is $RT = 0.082 \times 1500 = 123\ \ell$. Accordingly, the partition function at 1 mole ℓ^{-1} is

$$\frac{6.06 \times 10^{12}}{123} = 4.9 \times 10^{10}$$

This value may be compared with the value of 5.1×10^{10} calculated from molecular parameters, the difference probably being due to small errors in the many arithmetic steps involved, and perhaps small differences between our molecular parameters and those used in preparing the JANAF tables.

3.4.7 Sources of Molecular Data

Several collections of data available in most university libraries are very helpful in the assembly of the necessary molecular data for partition function calculations. Sutton (6) has made two extensive surveys of interatomic distances and molecular shapes, while several books, including Herzberg (7, 8) Nakamoto (9), and Wilson *et al.* (10), contain information on rotational and vibrational energy levels in molecules.

One problem that must be faced is the estimation of the properties of bonds of fractional order that form briefly during the lifetime of the activated complex. For example, in the reaction

$$D + H_2 \rightarrow D\text{—}H\text{—}H \rightarrow DH + H$$

it is clear that in the complex three electrons are responsible for forming two chemical bonds, each with a bond order of $\frac{3}{4}$, in simplest terms. Pauling (11) has suggested an empirical relationship between bond order and bond length

$$d_n = d_1 - 0.6 \log n \tag{3.60}$$

where d is a bond length in angstroms, n the bond order, and d_1 the length of a single bond. This is based on the observed interatomic distances in series such as C—C, C═C, and C≡C bonds, but there has been little opportunity to test it for bonds with $n < 1$ because such bonds are not stable. If the relatively short extension downward to $\frac{3}{4}$ is valid, we would write

$$d_{3/4} = d_1 - 0.6 \log \tfrac{3}{4} = d_1 + 0.08$$

Theoretical calculations on the hydrogen exchange reaction (12) indicate that the H—H distance in the complex is in the range 0.90–0.95 Å, compared to 0.74 Å in the H_2 molecule. The change is considerably more than 0.08 Å, and it seems better to take a $\frac{3}{4}$-order bond to be about 25% longer than a single bond, while recognizing that our lack of knowledge introduces possible error into the calculation.

For dissociation and association reactions, a method for calculating the length of the bond being broken or formed in the activated complex exists, and has recently been discussed by Waage and Rabinovitch (13).

An analogous problem arises when one must estimate vibrational frequencies for molecules with weak bonds. One useful relationship is that for any harmonic oscillator

$$\nu = \frac{1}{2\pi}\left(\frac{k_F}{\mu}\right)^{1/2} \tag{3.61}$$

where k_F is the force constant for the vibration, and μ is the reduced mass of the oscillating particles. Force constants for many molecules are given in the references on vibrational spectroscopy mentioned above. It turns out that there is an approximate correlation between force constant and bond strength, as shown in Table 3.3 and Fig. 3.3. An estimate of the bond strength may be obtained from bond strengths in molecules (14, 15) and an experimental or calculated value for the activation energy, while μ may be obtained from the geometry assumed for the complex.

While normal vibrations are properties of the molecule as a whole, many of these vibrations can be related substantially to particular bonds in the molecule, and we commonly refer to "C—H stretching vibrations," and so on. Because of this, it is usually safe to assume that vibrations that involve primarily parts of the molecule not directly involved in the reaction retain

Table 3.3 Correlation of Force Constant with Bond Energy

Bond	Bond energy (kcal)	Force constant (dyne/cm × 10^5)
N≡N	226	23.0
C≡C	194	16.4
C=O	170	12.6
C=C	147	9.7
H—F	135	9.7
H—O	111	7.8
H—Cl	103	4.8
C—H	99	4.8
H—Br	88	4.1
C—O	84	5.4
C—C	83	5.0
H—S	81	4.3
C—Cl	78	3.4
H—I	71	3.2

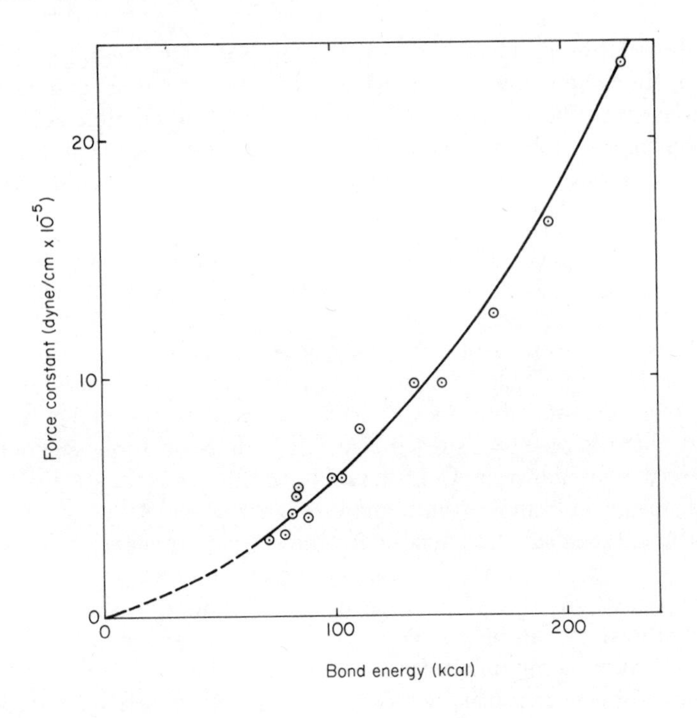

Fig. 3.3 Graph showing approximate relationship between bond energy and force constant.

the same frequencies in the complex as in the molecules. In this way we can limit our estimation of complex frequencies to the few that are intimately involved in the reaction.

3.5 DERIVATION OF THE ACTIVATED COMPLEX THEORY

Let us first consider a bimolecular reaction involving an atom and a diatomic molecule, which we can symbolize as

$$A + BC \rightarrow ABC \rightarrow AB + C$$

The equilibrium constant for the formation of activated complex ABC from A and BC will be

$$K^{\ddagger} = \frac{[ABC]}{[A][BC]} \tag{3.62}$$

so that $[ABC] = K^{\ddagger}[A][BC]$. However, this value of [ABC] will apply only when there is a complete equilibrium involving the products AB and C as well, and in that case half the ABC complexes will be bringing about reaction in each direction. The concentration of complexes that are reacting in the forward direction is

$$[\mathrm{ABC}] = \tfrac{1}{2}K^{\ddagger}[\mathrm{A}][\mathrm{BC}] \tag{3.63}$$

We can use the statistical thermodynamic formulation to write

$$K^{\ddagger} = \frac{NQ_{\mathrm{ABC}}}{Q_{\mathrm{A}}Q_{\mathrm{BC}}}\exp\left(\frac{-E_0}{RT}\right) \tag{3.64}$$

where the Q's are partition functions, N the Avogadro number, and E_0 the zero-point energy difference in the reaction by which the complex is formed. The partition functions of the reactants will be easily obtained from known data about them, but that for the complex will be harder to know. For the moment let us look at its vibrational energy levels. If ABC is linear, it will have four vibrational fundamentals, with motions (7, 9) as shown in Fig. 3.4. The third fundamental, in which A and B come together and C moves away, leads to the desired reaction and is called the *reaction coordinate.* It is clear that this will not be an ordinary vibration since it will occur only once as the reaction occurs. Indeed, as Fig. 3.4 indicates, only half of a vibrational cycle is needed to bring about reaction in one direction. Since the ABC molecule is unstable in terms of this vibration, it may be assumed that the force constant and thereby the vibrational frequency, for this motion are low. Accordingly, the contribution of this vibration to the partition function for ABC will be $\mathbf{k}T/h\nu_3$. If we now assume that we can work

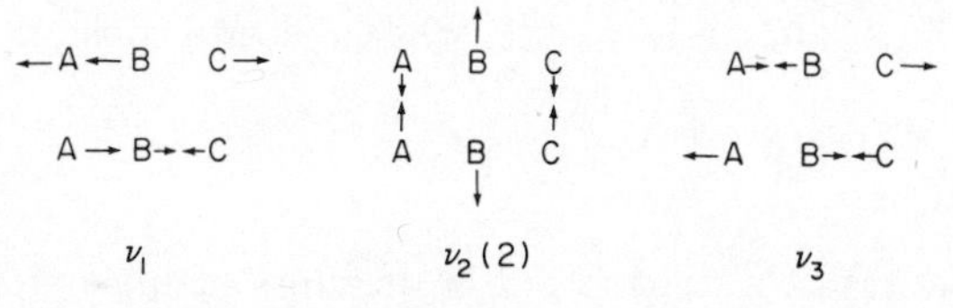

ν_1 $\nu_2(2)$ ν_3

Fig. 3.4 The three fundamental vibrations of a linear triatomic molecule. The second fundamental has a multiplicity of 2, the other motion being perpendicular to the paper. For each vibration, we have shown the motion during both halves of a vibration. The first half for ν_3 corresponds to the reaction A + BC → AB + C, while the second half corresponds to AB + C → A + BC. The relative degrees of motion will depend on the relative masses of the atoms. In ν_1, atom B will move in phase with the lighter of A and C, its displacement being less than that of its "partner." In a symmetrical case such as CO_2 or a H—H—H activated complex, the center atom will not move at all in ν_1, so that the stretching of the AB and BC bonds will be equal.

out the rest of the partition function somewhow, we will write

$$Q_{ABC} = Q_{\ddagger} \frac{\mathbf{k}T}{h\nu_3} \tag{3.65}$$

where $Q_{\ddagger}$ is the partition function for ABC *with the contribution of the reaction coordinate omitted.* Substitution of Eq. (3.64) gives

$$K^{\ddagger} = \frac{\mathbf{k}T}{h\nu_3} \frac{NQ_{\ddagger}}{Q_A Q_{BC}} \exp\left(\frac{-E_0}{RT}\right) \tag{3.66}$$

and substitution into Eq. (3.63) gives

$$[ABC] = \frac{1}{2} \frac{\mathbf{k}T}{h\nu_3} \frac{NQ_{\ddagger}}{Q_A Q_{BC}} \exp\left(\frac{-E_0}{RT}\right) [A][BC] \tag{3.67}$$

Of course, we cannot really calculate [ABC] since the frequency ν_3 is unknown. However, what we really want is the product of [ABC] and the rate constant for conversion of ABC into reaction products, and we can recognize that the time for a complex to dissociate will be $1/2\nu_3$, so the frequency (or rate constant) for decomposition will be $2\nu_3$. The rate of reaction is therefore

$$\text{rate} = 2\nu_3[ABC] \tag{3.68}$$

or

$$\text{rate} = \frac{\mathbf{k}T}{h} \frac{NQ_{\ddagger}}{Q_A Q_{BC}} \exp\left(\frac{-E_0}{RT}\right) [A][BC] \tag{3.69}$$

by substitution of Eq. (3.67) into (3.68), while the rate constant is

$$k = \frac{\mathbf{k}T}{h} \frac{NQ_{\ddagger}}{Q_A Q_{BC}} \exp\left(\frac{-E_0}{RT}\right) \tag{3.70}$$

In this equation the rate constant k will be in units of mole^{-1} ℓ sec^{-1} or mole^{-1} cc sec^{-1}, depending on the units chosen for the Q's. In many formulations of this equation N is omitted, in which case k would be in units of molecules rather than moles. In a unimolecular reaction in which a molecule A reacts, it is clear that a similar derivation would give

$$k = \frac{\mathbf{k}T}{h} \frac{Q_{\ddagger}}{Q_A} \exp\left(\frac{-E_0}{RT}\right) \tag{3.71}$$

in which the N's have completely canceled out.

The equation may also be derived by supposing that the reaction coordinate becomes a translational motion rather than a slow vibrational motion. It is assumed that the process occurs over a small distance δ, over which the average speed of the reacting molecules relative to one another is $\bar{c}$, which by the kinetic theory is given by

$$\bar{c} = \left(\frac{2\mathbf{k}T}{\pi\mu}\right)^{1/2} \tag{3.72}$$

where μ is the reduced mass of the particles. Accordingly, the time necessary for a complex to decompose is $\delta/\bar{c}$, or the rate is $\bar{c}/\delta$. In this instance the partition function for the reaction coordinate is, by Eq. (3.33),

$$\frac{(2\pi\mu\mathbf{k}T)^{1/2}\delta}{h} \tag{3.73}$$

so the expression for [ABC] becomes

$$[\mathrm{ABC}] = \frac{1}{2}\frac{(2\pi\mu\mathbf{k}T)^{1/2}}{h}\delta\frac{NQ_{\ddagger}}{Q_{\mathrm{A}}Q_{\mathrm{BC}}}\exp\left(\frac{-E_0}{RT}\right)[\mathrm{A}][\mathrm{BC}] \tag{3.74}$$

Combination of this equation with the rate of decomposition of complexes given above yields Eq. (3.69) for the rate and 3.70 for the rate constant.

It should be recognized that neither of the above derivations is rigorous. In the translation approach, δ is normally taken to represent a small distance—perhaps a few tenths of an angstrom—over which the largest part of the chemical reaction occurs. Macomber and Colvin (16) have pointed out that for light atoms, particularly hydrogen, a particle in this small a box may have widely spaced energy levels, while Eq. (3.73) would apply only if the energy levels are close. They propose that δ should be two mean free paths of the molecule, that is, the distance over which the total energy of the reacting molecules remains constant. This solves the problem of the size of the energy levels, except that μ changes as products are formed from reactants, and c may change considerably due to the changes in chemical potential energy as the reaction proceeds. Accordingly a more careful derivation would involve an integration over time rather than simple multiplication of terms. The results would be expected to be close to Eqs. (3.69) and (3.70), just as the careful analyses of Maxwell and Boltzmann give results differing by less than a factor of 2 from those of the simple kinetic theory.

3.5.1 Symmetry Numbers and Statistical Factors

We have noted above, in the discussion of rotational partition functions, that symmetry numbers enter into the partition functions and therefore into the rate constants. Since the symmetry numbers are in the denominators of the Q's, if we include them in the expression for the rate constant of a reaction

$$A + BC \rightarrow ABC \rightarrow ABC + C$$

then their contribution to the rate constant would be

$$\frac{\sigma_A \sigma_{BC}}{\sigma_{ABC}}$$

In many instances this symmetry effect is taken care of properly by the use of symmetry numbers. For example, for the reaction

$$H + CH_4 \rightarrow H{-}H{-}CH_3 \rightarrow CH_3 + H_2$$

we would consider that $\sigma_H = 1$, $\sigma_{CH_4} = 12$, and $\sigma_{CH_5} = 3$, so that

$$\frac{\sigma_A \sigma_{BC}}{\sigma_{ABC}} = \frac{1 \times 12}{3} = 4$$

It can be demonstrated that this is the right ratio by noting that, by symmetry, a reaction would occur if the H atom attacked any one of the four H atoms of methane, so that we would say the *statistical factor* is 4.

Schlag (17) and Bishop and Laidler (18) have pointed out some cases in which the symmetry number treatment gives different answers than the statistical factor method, the difficulty arising largely because ACT does not assume a true equilibrium concentration of complexes. It is concluded that a correct approach is to calculate the statistical factor by counting the number of ways in which the complex could be formed from the reactants, then using this number along with partition functions (Q') that do not include symmetry numbers.

3.6 A SAMPLE ACT CALCULATION

Let us calculate the frequency factor for the reaction

$$Br + Cl_2 \rightarrow BrCl + Cl$$

which has been studied in the gas phase at 293–333° by Christie *et al.* (19). They found

$$k = (4.5 \pm 2.0) \times 10^9 \, e^{-(6900 \pm 400)/RT} \quad \text{mole}^{-1}\,\ell\,\text{sec}^{-1}$$

The calculation, for a temperature of 300°K, involves several segments, as follows.

General Nature of the Activated Complex Probably a complex of the formula $BrCl_2$ will form, which will very likely be linear, as has been reported (9) for several negative ions of this general type. However, the complex will have a looser structure than the ions since only three electrons are available to form two bonds.

Statistical Factor Since the Br may react with either end of the Cl_2 molecule, the statistical factor is clearly 2. In this instance the symmetry numbers of the reactants (2 for Cl_2 and 1 for Br and $BrCl_2$) are such that they will give the same result as the statistical factor.

Partition Function for Br From the JANAF tables we find at 300°K

$$-\frac{F^\circ - H_{298}{}^\circ}{T} = 41.805 \text{ cal deg}^{-1}; \qquad H_0{}^\circ - H_{298}{}^\circ = -1.481 \text{ kcal}$$

from which $Q/N = 4.62 \times 10^6$ at 1 mole ℓ^{-1}.

Partition Function for Cl_2 From the JANAF tables we find at 300°K

$$-\frac{F^\circ - H_{298}{}^\circ}{T} = 53.289 \text{ cal deg}^{-1}; \qquad H_0{}^\circ - H_{298}{}^\circ = -2.194 \text{ kcal}$$

from which $Q/N = 4.53 \times 10^8$ at 1 mole ℓ^{-1}.

Partition Function for $BrCl_2$ This will have to be worked out piece by piece. For a molecular weight of $79.91 + 2 \times 35.45$ and a temperature of 300°K,

$$Q_T{}' = 0.311(300)^{3/2}(150.8)^{3/2} = 2.98 \times 10^6 \quad \text{at 1 mole } \ell^{-1}$$

According to Sutton's tables (6), single-bond Cl—Cl and Cl—Br distances are 1.99 and 2.14 Å. If it is assumed that in the compound these distances are 25% longer,[2] then we find $I = 1143 \times 10^{-40}$ g cm^2, and

$$Q_R = 0.0244 \times 1143 \times 300 = 8.37 \times 10^3$$

As mentioned above, several negative ions rather like our complex have been studied, the vibrational frequencies being listed by Nakamoto (Table 3.4). As in Section 3.5, ν_3 will be the reaction coordinate. There are numerous ways of estimating the most likely frequencies for the complex, given all these other ones for comparison. Rather than digress further at this point, let us choose two reasonable-looking frequencies, 190 cm^{-1} for ν_1 and 110 cm^{-1} for ν_2 (the doubly degenerate bending frequency) and go

[2] One may be as detailed as he wishes in these calculations, each extra consideration adding somewhat to the reliability of the final result. At this point it might be considered that the more electronegative chlorine might attract more than its share of electrons, so that the Cl—Cl bond might not be as much increased over its normal length as the Cl—Br bond. The changes, which would be only a few hundredths of an angstrom unit, would change I slightly.

Table 3.4

Compound	ν_1 (cm^{-1})	ν_2 (cm^{-1})	ν_3 (cm^{-1})
$(Br\ I\ Cl)^-$	198	135	180
$(Cl\ I\ Cl)^-$	268	129	218
$(I\ Br\ Br)^-$	143	98	178
$(Br\ I\ I)^-$	117	84	168
$(Br\ Br\ Br)^-$	170	—	210
$(Cl\ Cl\ Cl)^-$	268	165	242

ahead with the calculation. By Eq. (3.54) we find

$$Q_V = (1.67)(2.43)^2 = 9.8$$

Since there will be a considerable tendency for two of the bonding electrons to remain as a pair, we will estimate

$$Q_E = 4$$

which is the same as the ground state multiplicity for Br. Therefore, for the complex,

$$\frac{Q}{N} = (2.98 \times 10^6)(8.37 \times 10^3)(9.8)(4) = 9.8 \times 10^{11}$$

Calculation of the Rate Constant By the Activated Complex Theory (Eq. (3.70))

$$k = \frac{(1.38 \times 10^{-16})(300)}{6.63 \times 10^{-27}} \cdot \frac{9.8 \times 10^{11}}{(4.62 \times 10^6)(4.53 \times 10^8)} \exp\left(\frac{-E_0}{RT}\right)$$

$$= 8.6 \times 10^9 \exp\left(\frac{-E_0}{RT}\right) \quad \text{mole}^{-1}\ \ell\ \text{sec}^{-1}$$

Conversion to Arrhenius Form Since the temperature enters into the partition functions, the above equation is really of the form

$$BT^n \exp\left(\frac{-E_0}{RT}\right)$$

and we need to determine n in order to use the method of Eq. (3.12) to get A; that is

$$A = e^n BT^n$$

The value of n may be found by looking at the temperature dependences of the parts of the equation. To begin with, the $\mathbf{k}T/h$ term contributes one power of T. Q for Br contains $T^{3/2}$, and appears in the denominator. Q for Cl_2 contains $T^{3/2}$ for translation, T^1 for rotation, and perhaps there is some contribution from the vibrational partition function. To check on this last point, we note that the vibrational frequency for Cl_2 is 1.070 at 300°K and 1.120 at 360°K, a factor of 1.046 in Q_V for a factor of 1.2 in temperature, giving at $T^{0.25}$ dependence. Accordingly, Q for Cl_2 has a total temperature power of 2.75, appearing in the denominator. Q for $BrCl_2$ contains $T^{3/2}$ for translation, T^1 for rotation, and a total of $T^{2.25}$ for vibration, using the method suggested above for Cl_2, so its total power in temperature is 4.75. Accordingly,

$$\begin{aligned} n &= 1.00 + 4.75 - (1.50 + 2.75) = 1.50 \\ A &= e^{1.50} \times 8.6 \times 10^9 \\ &= 3.8 \times 10^{10} \quad \text{mole}^{-1}\, \ell\, \text{sec}^{-1} \end{aligned}$$

which is a factor of 8 higher than the experimental value of 4.5×10^9 mole^{-1} ℓ sec^{-1}.

Uses of the Calculation On the one hand, we could feel that the values are not in bad agreement, considering the estimates necessary in calculating the ACT value. The experimental value has been shown to be in a reasonable range, and it is unlikely that a gross error has been made in the determination.

On the other hand, if we had made our ACT calculation carefully and were now interested in extrapolating to obtain the rate constant at 500°K, we could use the calculation to improve the data. We note that the data were obtained over a fairly narrow temperature range, so that, even if the absolute values of the rate constants are quite good, the slope and intercept could be off somewhat. Accordingly, if we set $A = 4 \times 10^{10}$ and increase E from 6900 to 8700 cal, we obtain an equation that matches the ACT calculation and gives the experimental rate constant at 300°K. Of course, any equation with A somewhere between 4.5×10^9 and 4×10^{10}, with an appropriately chosen E, could be used to reflect our degrees of confidence in the observed and calculated values.

3.7 ESTIMATION OF ACTIVATION ENERGIES

While, as has been mentioned, there is no simple and accurate theory of activation energies, we are not left completely in the dark as to their probable values. Continuing to focus on elementary reactions, we would note that for an endothermic reaction the activation energy must be

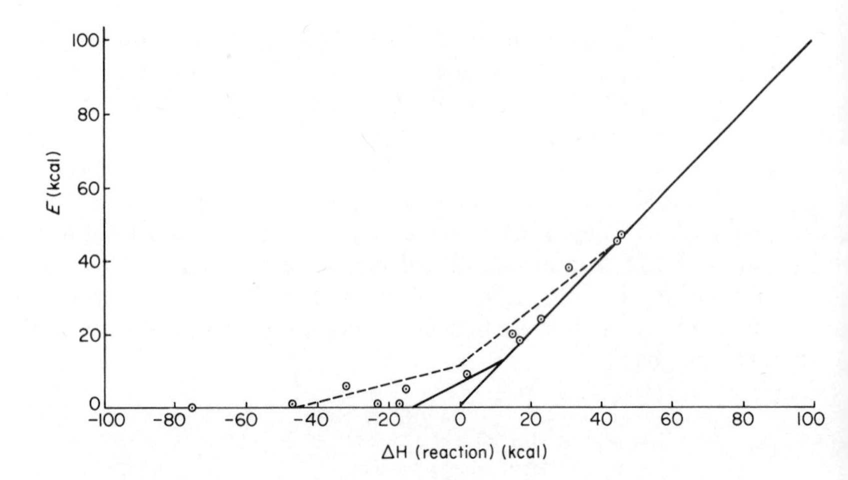

Fig. 3.5 Approximate relation between activation energy and energy of reaction for bimolecular exchange reactions.

at least equal to the energy of the reaction (with minor adjustments, such as we made in the last section, for temperature-dependent terms in the pre-exponential factor). Hammond (20) has generalized that for highly endothermic reactions the energy of the transition state is close to that of the products, while conversely for highly exothermic reactions it is close to that of the reactants. In the limit, E would be equal to ΔE for the reaction for endothermic reactions, and 0 for exothermic reactions, as shown schematically by the two lower straight lines in Fig. 3.4. These lines will tend to be asymptotes toward which activation energies will tend for highly endothermic or exothermic reactions. In the middle range, Kleinpaul (21) has suggested that E will seldom be less than $(7 + \frac{1}{2}\Delta E)$ kcal, and this line has also been drawn in Fig. 3.5. That these relationships are followed approximately for a considerable number of simple reactions is shown when E is plotted versus ΔH for the reactions of Table 3.1.

The dashed lines on the figure are the relationships suggested by Semenov (24). For $\Delta H < 0$,

$$E = 11.5 + 0.25\ \Delta H$$

while for $\Delta H > 0$

$$E = 11.5 + 0.75\ \Delta H$$

His relationships are dased on data for a variety of abstraction reactions rather like those of Table 3.1. It seems that if ΔH is known for a reaction of this type, E may be estimated within about 3 kcal.

The situation is entirely different for rearrangement reactions such as

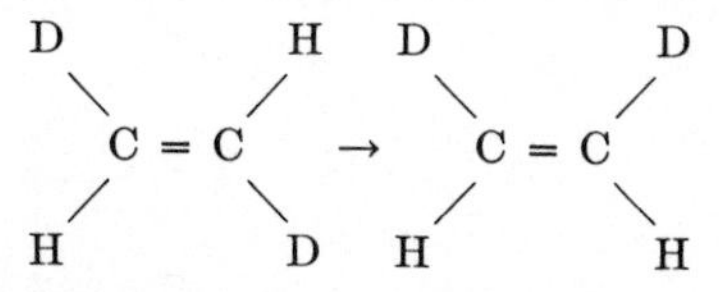

Here ΔH is small, but the amount of energy needed to loosen the bond to permit rearrangement is large—about 65 kcal—and there is no simple way to predict the amount. About the best one can do is to look for a reaction with a similar reaction path that has been measured. The systematic comparison of groups of solution reactions has been carried out extensively by use of equations due to Hammett and Taft. We will look into these further in Chapter 5.

References

1. A. L. Myerson and W. S. Watt, *J. Chem. Phys.* **49**, (1968).
2. W. C. Gardiner, Jr., "Rates and Mechanisms of Chemical Reactions," pp. 78–100. Benjamin, New York, 1969.
3. H. Eyring, *J. Chem. Phys.* **3**, 107 (1935).
4. D. R. Stull and H. Prophet, "JANAF Thermochemical Tables," 2nd ed. U.S. Government Printing Office, Washington, D.C., 1971.
5. Selected Values of Properties of Hydrocarbons and Related Compounds. American Petroleum Institute Research Project 44. Thermodynamics Research Center, Department of Chemistry, Texas A and M University, College Station, Texas.
6. L. E. Sutton, "Tables of Interatomic Distances and Configuration in Molecules and Ions." Special Publications Nos. 11 and 18. The Chemical Society, London, 1958 and 1965.
7. G. Herzberg, "Molecular Spectra and Molecular Structure. II. Infrared and Raman Spectra of Polyatomic Molecules." Van Nostrand, Princeton, New Jersey, 1945.
8. G. Herzberg, "The Spectra and Structures of Simple Free Radicals." Cornell Univ. Press, Ithaca, New York, 1971.
9. K. Nakamoto, "Infrared Spectra of Inorganic and Co-ordination Compounds," 2d Ed., Wiley (Interscience), New York, 1970.
10. E. B. Wilson, Jr., J. C. Decius, and P. C. Cross, "Molecular Vibrations." McGraw-Hill, New York, 1955.
11. L. Pauling, *J. Amer. Chem. Soc.* **69**, 542 (1947).
12. R. E. Weston, Jr., Science **158**, 332 (1967).
13. E. V. Waage and B. S. Rabinovitch, *Chem. Rev.* **70**, 377 (1970).
14. S. W. Benson, "Thermochemical Kinetics." Wiley, New York, 1968.
15. A. G. Gaydon, "Dissociation Energies." 3d Ed.. Chapman and Hall, London, 1968.
16. J. D. Macomber and C. Colvin, *Int. J. Chem. Kinet.* **1**, 483 (1969).
17. E. W. Schlag, *J. Chem. Phys.* **38**, 2480 (1963).
18. D. M. Bishop and K. J. Laidler, *J. Chem. Phys.* **42**, 1688 (1965).

19. M. I. Christie, R. S. Roy, and B. A. Thrush, *Trans. Faraday Soc.* **55,** 1139 (1959).
20. G. S. Hammond, *J. Amer. Chem. Soc.* **77,** 334 (1955).
21. W. Kleinpaul, *Z. Phys. Chem.* **26,** 313 (1960); **27,** 343 (1961); **29,** 201 (1961).
22. N. N. Semenov, "Some Problems in Chemical Kinetics and Reactivity." Pergamon Press, London and New York, 1958.

Further Reading

S. W. Benson, "Thermochemical Kinetics. Methods for the Estimation of Thermochemical Data and Rate Parameters." Wiley, New York, 1968 (a very practical reference for the kind of estimates needed for ACT calculations).

S. Glasstone, K. J. Laidler, and H. Eyring, "The Theory of Rate Processes." McGraw-Hill, New York, 1941 (masterful presentation of ACT, and much more, that is still a definitive text after a third of a century).

J. O. Hirschfelder and D. Henderson, eds., "Chemical Dynamics. Papers in Honor of Henry Eyring." Wiley (Interscience), New York, 1971 (state-of-the-art papers by experts).

R. D. Levine, "Quantum Mechanics of Molecular Rate Processes." Clarendon Press, Oxford, 1971 (very good, very difficult).

O. K. Rice, "Statistical Mechanics, Thermodynamics and Kinetics." Freeman, San Francisco, 1967 (a unique integration of these three subject areas).

Problems

3.1 Show in detail how substitution of Eq. (3.4) into Eq. (3.5) leads to Eq. (3.6).

3.2 Use the collision theory to calculate the preexponential factor for one of the reactions of Table 3.1, or of some other reaction for which you have experimental data for comparison. A good source for collision diameters is "Molecular Theory of Gases and Liquids," by J. O. Hirschfelder, C. F. Curtiss, and R. B. Bird (New York, Wiley, 1954).

3.3 Calculate the fraction of collisions that have energies greater than 100, 1000, and 10,000 cal/mole at 300°K. How many such collisions would there be per second in 1 cc of N_2 at 1 atm pressure?

3.4 According to the variation of the collision theory described at the end of Section 3.1, the quantity σ_{12}^2 in Eq. (3.10) is replaced by the reaction cross section, σ_R^2. Using the data of Table 3.1, calculate σ_R for the reactions

$$HO_2 + Ar \rightarrow H + O_2 + Ar$$

and

$$CO + OH \rightarrow CO_2 + H$$

which appear to have A values above and below the average. Compare these σ_R values with the expected kinetic theory σ_{12} quantities.

3.5 Suppose that two energy levels in a molecule differ in energy by 1×10^{-13} erg. Express this energy difference in terms of cm^{-1}, cal/mole and joule/mole. If the multiplicities of both levels are 1, calculate the ratio of the number of molecules in each at 100°, 500°, 1000°, and 3000°K.

3.6 Suppose we had some imaginary molecules of type A, which had just 3 energy levels of multiplicity 1, spaced 500 cm^{-1} apart, and some other molecules of type B, which had 2 energy levels of multiplicity 3, spaced 700 cm^{-1} apart, the lower energy of type B being 300 cm^{-1} higher than the lowest energy of type A. Calculate the equilibrium constant for the reaction $A \rightleftarrows B$ and the fraction of each type of molecule present at 100 and 1000°K. What would be the limiting values of the equilibrium constant and the fractions at 0°K and at infinite temperature?

3.7 Calculate the partition function of some molecule at 500°K using molecular data available in the JANAF tables or elsewhere. Check your number by calculation of the partition function from tabulated thermodynamic data. Be sure to specify the concentration units, and make the comparison in the same units.

3.8 In Section 3.3 it is stated that the method of obtaining the total partition function of a real molecule by combining the separate partition functions of various types of energy (Eq. (3.30)) is an approximation. What features of the motions of real molecules cause this method to be approximate? If corrections were made, would the accurate total partition function tend to be larger or smaller than the approximate one?

3.9 What would be the statistical factors for the following reactions?

(a) $H + CH_4 \rightarrow CH_3 + H_2$
(b) $OH^- + CHCl_3 \rightarrow CHCl_2OH + Cl^-$
(c) $D_2 + CH_3 \rightarrow CH_3D + D$
(d)

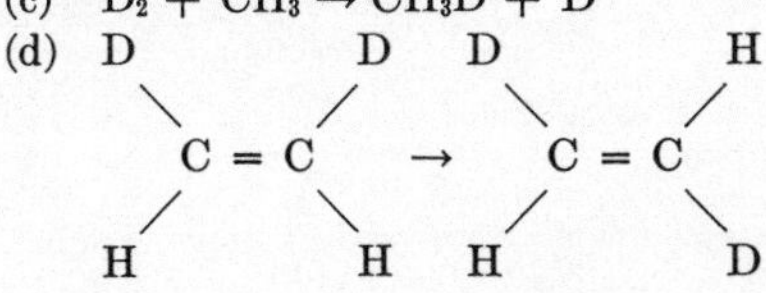

3.10 Estimate moments of inertia and vibrational frequencies for activated complexes formed in the following reactions:

(a) $O + OH \rightarrow (OOH^{\ddagger}) \rightarrow O_2 + H$
(b) $F + H_2 \rightarrow (FHH^{\ddagger}) \rightarrow HF + H$
(c) $N_2O + O \rightarrow (ONNO^{\ddagger}) \rightarrow NO + NO$

4 Some Typical Gas-phase Reactions

In this and the next two chapters we will look at examples of reactions in the gas, liquid, and solid phases, including some heterogeneous reactions.

4.1 THE VARIABLE REACTION ORDER OF UNIMOLECULAR REACTIONS

It has been found that a large number of apparently elementary reactions involving a single molecule display first-order kinetics at high pressures and second-order kinetics at low pressures, with a gradual transition between the two regions. The data are often expressed in terms of a "fall-off" of the first-order rate constant with decreasing pressure since that is an easy way to visualize them. Figure 4.1 illustrates a "fall-off curve" for the conversion of methyl isocyanide to acetonitrile, as studied by Schneider and Rabinovitch (1). It is conventional to call the limiting first-order rate constant at high pressure k_∞. As the pressure is lowered, the observed rate of reaction is converted into a first-order rate constant, which takes on steadily lower values. Eventually, as the reaction becomes second order, the "first-order rate constant" becomes directly proportional to the pressure. Some other reactions that behave this way are the cis–trans isomerization of a number of olefinic compounds such as *cis*-2-butene and *cis*-ethylene-d_2, and the decomposition of hydrocarbons such as methane and ethane. The phenomenon is really quite common and often encountered. Each reaction has a characteristic pressure at which the fall-off occurs—in general, the smaller the molecule and the higher the temperature, the higher the fall-off pressure.

Interestingly, the explanation was put forward (in a discussion on the driving force leading to chemical reactions) before the phenomenon itself was observed (2). It is assumed that a unimolecular reaction occurs in two stages,[1] which may be symbolized as follows:

$$A + A \underset{-1}{\overset{1}{\rightleftarrows}} A^* + A$$

$$A^* \xrightarrow{2} \text{products}$$

According to the first equation, from time to time a collision between two molecules will leave one of them (A^*) with sufficient energy for reaction. The energized molecule may either react by reaction 2, or lose its energy by reaction −1. In a given situation, a steady state concentration of energized molecules will result; this can be calculated by setting the net rate of formation of A^* molecules to zero:

$$\frac{d[A^*]}{dt} = k_1[A]^2 - k_{-1}[A^*] - k_2[A^*] = 0$$

If we solve this equation for $[A^*]$, the result is

$$[A^*] = \frac{k_1[A]^2}{k_{-1}[A] + k_2} \tag{4.1}$$

and the rate of reaction (that is, the rate of formation of products, and therefore the net rate at which A disappears) is

$$k_2[A^*] = \frac{k_1k_2[A]^2}{k_{-1}[A] + k_2} \tag{4.2}$$

Two limiting cases of Eq. (4.2) may be seen. If the pressure is high, then $k_{-1}[A]$ will be greater than k_2, and will dominate in the denominator, so

[1] In some treatments three steps are written, by breaking up the second into

$$A^* \rightarrow A^\ddagger$$

$$A^\ddagger \rightarrow \text{products}$$

where $A^\ddagger$ is the activated complex. However, by this time we can recognize that every reaction must proceed by an activated complex, so that it does not seem necessary to write specifically that it will exist for this reaction since no discussion of its role or nature is included in the simple Lindemann hypothesis.

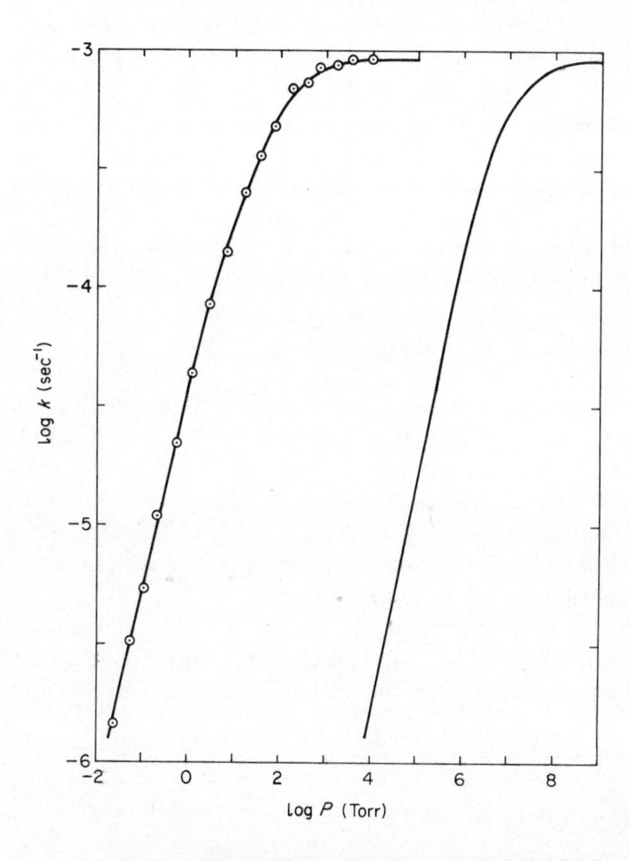

Fig. 4.1 First-order rate constant for thermal isomerization of methyl isocyanide as a function of pressure at 230.4°C. -O-O- Data of Schneider and Rabinovitch (1). Curve on right found by simple Lindemann theory (Problem 4.1).

that

$$\text{rate} = \frac{k_1 k_2}{k_{-1}}[\mathrm{A}] \qquad \text{or} \qquad k_\infty = \frac{k_1 k_2}{k_{-1}} \tag{4.3}$$

and the reaction is first order at high pressure. On the other hand, at low pressure, $k_{-1}[\mathrm{A}]$ will be small compared to k_2, so

$$\text{rate} = k_1[\mathrm{A}]^2 \qquad \text{or} \qquad k_{\text{first-order}} = k_1[\mathrm{A}] \tag{4.4}$$

and the reaction is second order at low pressures, with the apparent first-order rate constant being proportional to the concentration (or pressure). At intermediate pressures, reaction orders between 1 and 2 are indicated by Eq. (4.2), with the apparent first-order rate constant given by

$$k_{\text{first-order}} = \frac{k_1 k_2[\mathrm{A}]}{k_{-1}[\mathrm{A}] + k_2} \tag{4.5}$$

The Lindemann proposal, therefore, gives the correct general form of the fall-off curve.

From our earlier considerations we can estimate the values of k_1 and k_{-1}. They are simply collision theory rate constants, so that by Eq. (3.13)

$$k_1 = 2N\sigma^2\left(\frac{\pi \mathbf{k}T}{m}\right)^{1/2} \exp\left(\frac{-E_{\mathrm{CT}}}{RT}\right) \tag{4.6}$$

where E_{CT} is the energy needed for reaction. The expression for k_{-1} will be identical except that the exponential term will be e^0 or 1 since essentially all collisions of a highly energized molecule will result in loss of energy to the point where the molecule can no longer react. (This is sometimes called the "strong collision" assumption.) The ratio k_1/k_{-1}, which is $\exp(-E_{\mathrm{CT}}/RT)$, is the fraction of molecules having energy greater than E_{CT}, as given by the Boltzmann distribution. We can also recognize that k_2 will be a frequency analogous to the ACT frequency ν at which energized molecules react to products. Another look at Eq. (4.3) suggests that if we have an Arrhenius equation for a unimolecular reaction at high pressure, then we can write

$$\mathrm{A} \exp\left(\frac{-E}{RT}\right) = k_2 \exp\left(\frac{-E_{\mathrm{CT}}}{RT}\right) \tag{4.7}$$

so that, by Eqs. (3.11) and (3.12),

$$k_2 = A/e^{1/2} \qquad \text{and} \qquad E_{\mathrm{CT}} = E - \tfrac{1}{2}RT \tag{4.8}$$

This analysis (which should be followed through numerically by doing Problem 4.1) leads to the right-hand fall-off curve of Fig. 4.1, which is qualitatively correct but quantitatively different from experiment by a large amount.

An extension of the Lindemann idea that gives a much more adequate description of unimolecular reactions has been made, called the RRKM theory after the initials of its propounders (3–5). Lindemann's two-step process is made more specific by the postulate that only the energy in the form of vibration and internal rotations can be used for reaction.[2] Reaction

[2] This statement is not quite true for two reasons. First, it is usually assumed that the overall rotations of the molecule remain in the same quantum states throughout the reaction, and since in the activated complex the moments of inertia are usually different (typically larger) than in the molecule, some energy may be contributed to the reaction by this change (6). Second, in a few cases, one or more rotational degrees of freedom may contribute significantly to the reaction process.

occurs when a sufficient amount of energy happens to be (by statistical probability) transferred into the reaction coordinate.

Moreover, the rate at which a molecule will react is taken to depend on the amount of energy it has. The rate constant for the conversion of energized molecules to activated complexes, and thence to products, as a function of energy, formerly called k_2 but now symbolized as k_E, is calculated by a statistical method very like that of the ACT. The "rate constant" for the reaction will then be the sum of the contributions at different energies.

Let us rewrite Eq. (4.5) in this new form:

$$k_{\text{first order}} = \int_{E_0}^{\infty} \frac{k_1 k_E [\text{A}]\, dE}{k_{-1}[\text{A}] + k_E} \tag{4.9}$$

Naturally, molecules must have energy at least equal to E_0, the zero-point energy of reaction, to react. Let us rearrange this equation by dividing top and bottom by $k_{-1}[\text{A}]$

$$k_{\text{first order}} = \int_{E_0}^{\infty} \frac{(k_1 k_E / k_{-1})}{1 + (k_E / k_{-1}[\text{A}])}\, dE \tag{4.10}$$

and then replacing the ratio k_1/k_{-1} in the numerator by the Boltzmann distribution expression $B(E)$

$$k_{\text{first order}} = \int_{E_0}^{\infty} \frac{B(E) k_E\, dE}{1 + (k_E / k_{-1}[\text{A}])} \tag{4.11}$$

It is important to note that k_1 and, therefore, the Boltzmann ratio k_1/k_{-1}, do not have the same values as in the simple Lindemann mechanism since we are now allowing the molecules to accumulate energy in internal degrees of freedom, rather than having to convert it directly from collisional energy to the reaction coordinate. However, we will retain the assumption that an energized molecule has a 100% probability of becoming deenergized on collision.

The calculation of $B(E)$ and k_E is often practically difficult for actual molecules, but the methods of obtaining them are not difficult in principle. Let us look at each one in turn.

$B(E)$, in this calculation, is the fraction of molecules that have internal (or "active") energies of a certain value under conditions of thermal equilibrium—that is, at high pressure. Specifically, if we have N_0 molecules altogether, then

$$B(E) = \frac{1}{N_0} \frac{dN}{dE} \tag{4.12}$$

where in molecular terms, E could be in ergs. Once we have decided which energy levels we wish to include in the calculation, it is possible to calculate the density of states, or the number of "active" quantum states per erg, at any given energy. This number is symbolized by $N(E)$. In a small increment of energy dE, the number of molecules will be proportional to $N(E)e^{-E/RT}$, while the sum (or integral) of all such terms, at all energies will be the partition function Q_{PD}.[3] Therefore,

$$B(E) = \frac{N(E)e^{-E/RT}}{Q_{PD}} \tag{4.13}$$

and, of course,

$$\int_0^\infty B(E)\, dE = 1 \tag{4.14}$$

As a simple example, suppose we consider a single vibrational degree of freedom, with $\tilde{\nu} = 300\ \text{cm}^{-1}$. This corresponds to a quantum state spacing of $300hc = 0.595 \times 10^{-13}$ erg, so $N(E) = 1/0.595 \times 10^{13} = 1.68 \times 10^{13}\ \text{erg}^{-1}$ at all energies.[4] If we assume a temperature of 300°K, then by Eq. (3.54), $Q_{PD} = 1.31$, so $B = 1.28 \times 10^{13}\ e^{-E/RT}\ \text{erg}^{-1}$. To check Eq. (4.14) by integrating, we must use consistent units (that is, E in ergs and $\mathbf{k}$ in place of R) which gives

$$\int_0^\infty B(E)\, dE = (1.28 \times 10^{13})(\mathbf{k}T)$$

$$= (1.28 \times 10^{13})(1.38 \times 10^{-16})(300) = 0.53$$

The reason we did not get 1 is that so few quantum states are involved that the integral is not an adequate representation of what is really a quantum-mechanical sum. If we try 3000°K, at which $Q_{PD} = 7.46$, then $B = 2.25 \times 10^{12}\ e^{-E/RT}\ \text{erg}^{-1}$, and

$$\int_0^\infty B(E)\, dE = 0.93$$

We will clearly have to be careful in mixing quantum and classical statistics in our calculations. Fortunately, the facts that we will usually have quite a

[3] The subscript on Q_{PD} is taken to imply the partition function for those participating degrees of freedom that contribute energy toward reaction, normally the molecular vibrations and internal rotations.

[4] We could also have expressed $N(E)$ as 0.0033 levels/cm^{-1}, or as $1.68 \times 10^{20}\ \text{J}^{-1}$. Probably because vibrational energy level spacings are most commonly given in cm^{-1}, most RRKM calculations in the literature use this unit.

few degrees of freedom participating, and that we will be considering quite high energies in order to bring about reaction, will minimize the difficulties.

With systems of several degrees of freedom, $N(E)$ increases rapidly with increasing energy. For example, let us look at a system with three harmonic vibrational degrees of freedom, with spacings of 200, 300, and 400 cm^{-1}. The method of counting the number of quantum states (8, 9) at each energy is illustrated in Table 4.1. In the second column we have listed the multiplicities to be expected from the 200-cm^{-1} frequency alone (a multiplicity of 1 every 200 cm^{-1}). In the next column we have included the effect of the 300-cm^{-1} frequency, and in the fourth column, the effect of the 400-cm^{-1} frequency. Formally, one can generate the number at any location in the table by adding together the multiplicity in the space to the left (which is the multiplicity due to frequencies that have been considered previously) and that of the space with energy of one quantum less than the one in question in the same column (which is the contribution to the multiplicity

Table 4.1 Enumeration of the Number of States at Lower Energies for an Illustrative Example of a Molecule with Three Vibrational Frequencies of 200, 300, and 400 cm^{-1}

	Multiplicity due to			
Energy (cm^{-1})	200 cm^{-1}	300 cm^{-1}	400 cm^{-1}	
2000	1	4	14	
1900	0	3	10	
1800	1	4	12	
1700	0	3	8	Total of 85
1600	1	3	10	
1500	0	3	7	Average $N(E)$
1400	1	3	8	0.085/cm^{-1}
1300	0	2	5	
1200	1	3	7	
1100	0	2	4	
1000	1	2	5	
900	0	2	3	
800	1	2	4	Total of 23
700	0	1	2	
600	1	2	3	Average $N(E)$
500	0	1	1	0.023/cm^{-1}
400	1	1	2	
300	0	1	1	
200	1	1	1	
100	0	0	0	
0	1	1	1	

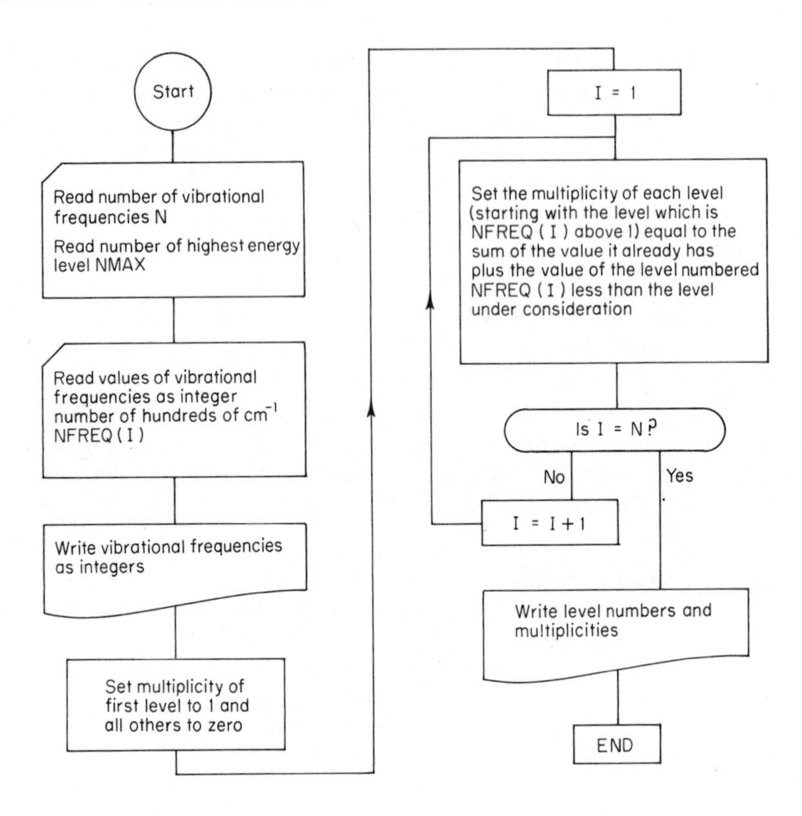

Fig. 4.2 Logic diagram for a FORTRAN program for calculating the number of vibrational quantum states at a given energy. For this illustrative program, the storage needed in the computer has been kept to a minimum by rounding all frequencies to the nearest 100 cm, and tabulating multiplicites at energy increments of 100 cm. The zero energy level is numbered 1 in the program, the energy level at 100 cm^{-1} is numbered 2, and so on.

of the frequency being considered). For example, the 2 in the 300-cm^{-1} column at an energy of 600 cm^{-1} comes about by adding the 1 in the 200-cm^{-1} column at 600-cm^{-1} energy and the 1 in the 300-cm^{-1} column at 300-cm^{-1} energy. We are saying that a 600-cm^{-1} energy can be obtained either by 3 quanta of 200 cm^{-1} each or by 2 quanta of 300 cm^{-1} each. When we add the third frequency, we also include the possibility of 1 quantum of 200 cm^{-1} plus 1 of 400 cm^{-1} for a total multiplicity of 3.

A computer program is clearly needed to carry out calculations over a practical range. In Fig. 4.2 we describe a simple one that will reproduce the data of Table 4.1, and go on to substantially higher energies. For example, there are 867 combinations of quantum states that give a total energy of 19,800 cm^{-1} in the example of Table 4.1.

The value of k_E to be used in Eq. (4.11) is given approximately by the equation

$$k_E = \frac{m}{h} \frac{Z^{\ddagger}}{Z} \frac{G^{\ddagger}(E)}{N(E)} \tag{4.15}$$

where $N(E)$ is the density of states of the reacting molecule, discussed above, in erg^{-1}, $G^{\ddagger}(E)$ is the sum of all the participating quantum states in the activated complex up to the energy E, $Z^{\ddagger}$ and Z are the moment of inertia terms in the rotational partition functions of the activated complex and molecule (see Eqs. (3.48) and (3.49)), h is Planck's constant, and m is the statistical factor for the reaction. The concept behind this equation is that molecules with a given internal energy can convert to complexes with that internal energy, *and also* to complexes with any lower allowed internal energy, the balance of the energy being put into the reaction coordinate (for example, translational energy for a dissociation, or rotation for a cis–trans isomerization). This is illustrated in Fig. 4.3. Planck's constant appears as a proportionality constant to convert k_E into units of inverse seconds. As previously stated, k_E is calculated strictly from the number of states available in the molecule and the complex at a given energy. We should note that while N is a density of states and G is a sum,

Table 4.2 A FORTRAN Program for Calculating the Number of Harmonic Vibrational Quantum States at a Given Energy

```
C      PROGRAM COUNT
       DIMENSION MULT(1000),NFREQ(10)
       READ(5,100)N,NMAX
  100  FORMAT(2I5)
       READ(5,101)(NFREQ(I),I=1,N)
  101  FORMAT(10I5)
       WRITE(6,200)(NFREQ(I),I=1,N)
  200  FORMAT(1H1,2X,'CALCULATION OF MULTIPLICITIES OF ENERGY LEVELS',//,
      1 3X,'FREQUENCIES (/100) ARE   ',10I5,///)
       MULT(1) = 1
       DO 10 J=2,NMAX
       MULT(J) = 0
   10  CONTINUE
       DO 20 I=1,N
       NDIM = NMAX - NFREQ(I)
       DO 30 L=1,NDIM
       KK = L + NFREQ(I)
       MULT(KK) = MULT(KK) + MULT(L)
   30  CONTINUE
   20  CONTINUE
       WRITE(6,201)(J,MULT(J),J=1,NMAX)
  201  FORMAT(/,2X,10(2I6))
       STOP
       END
```

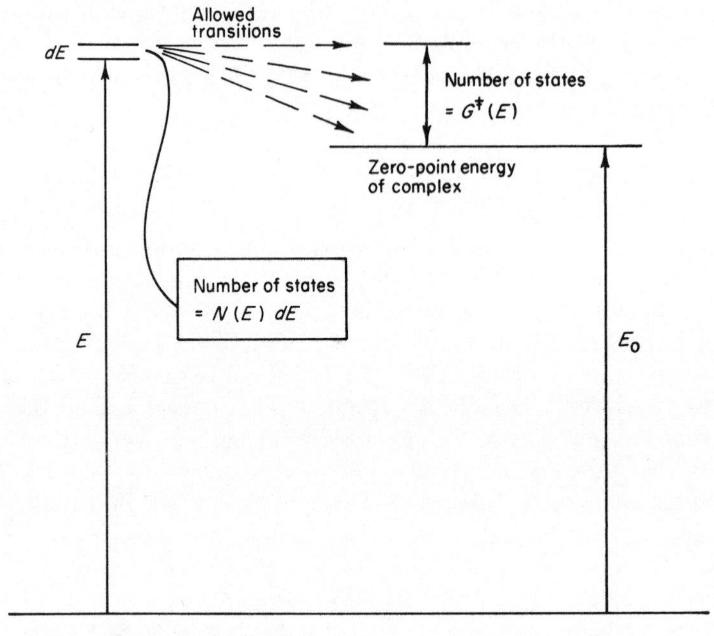

Fig. 4.3 Illustration of the method of obtaining k_E in the RRKM theory.

the number of states in a small range dE of a few cm^{-1} for the energized molecule may be greater than the number in 1000 or more cm^{-1} for the complex since, as we have seen, the density of states is low at low energies (relative to the zero-point energy).

Calculation of k_E involves making estimates of the nature of the activated complex, exactly as must be done to apply the ACT. One of the frequencies of the molecule will become the reaction coordinate, and usually some of the other frequencies will be expected to change their values. Sometimes internal rotations must be considered, perhaps being present in the complex if not in the molecule (as in a cis–trans isomerization). Less often, overall rotations of the molecule must be included, if, for instance, a linear complex is thought to form from a bent molecule, or vice versa. More usually, it is considered that stretching of a bond increases two of the moments of inertia of a molecule, while leaving the other constant. For example, in the dissociation of ethane into two methyl groups the complex would probably have a longer C—C bond distance than the molecule, which would increase the two large moments of inertia, but leave unchanged the smaller one for rotation around the C—C axis. In this instance, since in the partition function the product of the moments of inertia is taken to the $\frac{1}{2}$ power, the ratio $Z^\ddagger/Z$ is simply the ratio of the two large moments of inertia.

Actually, the ACT and RRKM theory give identical results for k_∞. This can be seen if one obtains k_∞ from Eq. (4.1):

$$k_\infty = \int_{E_0}^{\infty} B(E) k_E \, dE \tag{4.16}$$

and then substitutes the values of $B(E)$ and k_E to obtain

$$k_\infty = \frac{m}{hQ_{\mathrm{PD}}} \frac{Z^\ddagger}{Z} \int_{E_0}^{\infty} G^\ddagger(E) \exp\left(\frac{-E_0}{RT}\right) dE \tag{4.17}$$

As Marcus pointed out, since

$$G^\ddagger(E) = \int_0^E N^\ddagger(E) \, dE \tag{4.18}$$

and since the energy E_0, measured from the zero energy state of the molecule, is 0 energy from the point of view of the complex, then

$$\int_{E_0}^{\infty} G^\ddagger(E) \exp\left(\frac{-E}{RT}\right) dE = kTQ_{\mathrm{PD}} \exp\left(\frac{-E_0}{RT}\right) \tag{4.19}$$

so that

$$k_\infty = \frac{m\mathbf{k}T}{h} \frac{Z^\ddagger}{Z} \frac{Q_{\mathrm{PD}}{}^\ddagger}{Q_{\mathrm{PD}}} \exp\left(\frac{-E_0}{RT}\right) \tag{4.20}$$

This is the same as Eq. (3.71) since the expression

$$\frac{Z^\ddagger}{Z} \frac{Q_{\mathrm{PD}}{}^\ddagger}{Q_{\mathrm{PD}}}$$

is the ratio of the partition functions of the activated complex (omitting the reaction coordinate) and the molecule, the translational partition functions, which are equal, having been canceled out, along with (usually) one of the rotational partition functions, for which the moment of inertia has been assumed not to have changed.

Finally, let us take a brief look at what the ACT and RRKM theory predict for the Arrhenius activation energies of unimolecular reactions. In general, the Arrhenius energy is given by

$$E = -R \frac{d \ln k}{d(1/T)} \tag{4.21}$$

a relationship that is implicit from Eq. (2.40). At high pressure, from Eq. (4.20),

$$\ln k_\infty = \ln \frac{mkZ^\ddagger}{hZ} + \ln T + \ln Q_{\mathrm{PD}}{}^\ddagger - \ln Q_{\mathrm{PD}} - \frac{E_0}{RT}$$

$$-\frac{d \ln k_\infty}{d(1/T)} = 0 + T - \frac{d \ln Q_{\mathrm{PD}}{}^\ddagger}{d(1/T)} + \frac{d \ln Q_{\mathrm{PD}}}{d(1/T)} + \frac{E_0}{R}$$

$$E = E_0 + RT - R\frac{d \ln Q_{\mathrm{PD}}{}^\ddagger}{d(1/T)} + R\frac{d \ln Q_{\mathrm{PD}}}{d(1/T)} \tag{4.22}$$

We may evaluate the last two terms for a particular reaction once we have specified what degrees of freedom are involved in the Q's. If they are all vibrations, for example, we know that the Q's consist of products of terms, each of the form $(1 - \exp(-E_{\mathrm{V}}/RT))^{-1}$, from Eq. (3.54), where E_{V} is the vibrational energy level spacing. The logarithms of the Q's will be sums of such terms, and when the derivatives are taken with respect to $1/T$ we obtain sums of terms so that, for example,

$$-R\frac{d \ln Q_{\mathrm{PD}}{}^\ddagger}{d(1/T)} = \frac{E_{\mathrm{V}1} \exp(-E_{\mathrm{V}1}/RT)}{1 - \exp(-E_{\mathrm{V}1}/RT)} + \frac{E_{\mathrm{V}2} \exp(-E_{\mathrm{V}2}/RT)}{1 - \exp(-E_{\mathrm{V}2}/RT)} + \cdots \tag{4.23}$$

Each of these terms approaches 0 for large E_{V} or small T, and approaches RT for small E_{V} or large T. Clearly, the Arrhenius activation energy will tend to change with temperature, although, since the changes in RT and in the vibrational contributions may be small compared to E_0 over a limited experimental range, the temperature dependence of E may not be conspicuous. Because the looser structure of the activated complex usually leads to some of its vibrational frequencies being lower than the corresponding ones of the molecule, it is common for E at high pressure to be several RT units larger than E_0.

A similar derivation for the Arrhenius activation energy at low pressure can be made starting with Eq. (4.11). At low pressure the apparent first-order rate constant becomes

$$k = k_{-1}[\mathrm{A}] \int_{E_0}^{\infty} B(E)\, dE \tag{4.24}$$

and if we assume constant pressure conditions, the Arrhenius activation energy E is

$$E = E_{av} - \frac{1}{2}RT + R\frac{d \ln Q_{PD}}{d(1/T)} \tag{4.25}$$

where E_{av} is the low pressure[5] average energy of the reacting molecules, given by

$$E_{av} = \frac{\int_{E_0}^{\infty} EN(E)\, e^{-E/RT}\, dE}{\int_{E_0}^{\infty} N(E)\, e^{-E/RT}\, dE} \tag{4.26}$$

This average energy, which is somewhat greater than E_0, can be obtained by a numerical integration once $N(E)$ has been found. The usual result of the calculation is that the Arrhenius activation energy at low pressure is less than that at high pressure, and may well be less than E_0. At intermediate pressures, of course, experimental E's between the two limiting values may be found.

4.2 FREE RADICALS IN GASEOUS REACTIONS

In many reactions very reactive, and therefore short-lived, intermediates are found to be present. As one of the earliest examples, it was proposed independently and almost simultaneously by Christiansen (10), K. F. Herzfeld (11), and Polanyi (12) about 1920 that H and Br atoms propagate the reaction between hydrogen and bromine vapor. At the beginning of the process three reactions occur:

$$Br_2 \overset{k_1}{\leftrightarrows} 2\,Br$$

$$Br + H_2 \xrightarrow{k_2} HBr + H$$

$$H + Br_2 \xrightarrow{k_3} HBr + Br$$

A small concentration of Br atoms is established, the value being given by

[5] At high pressure the average energy of the reacting molecules will be higher. This comes about because the fast-reacting high-energy molecules are replaced at infrequent intervals at low pressures, but frequently (instantly) at high pressure. Accordingly, the proportion of reacting molecules with high energy increases at high pressure.

the equilibrium expression

$$k_1 = \frac{[Br]^2}{[Br_2]}$$

so that

$$[Br] = k_1^{1/2}[Br_2]^{1/2}$$

The concentration of H atoms also reaches a steady value such that the rate of production by reaction 2 is equal to the loss by reaction 3:

$$\frac{d[H]}{dt} = k_2[Br][H_2] - k_3[H][Br_2] = 0$$

from which

$$[H] = \frac{k_2[H_2]}{k_3[Br_2]}[Br]$$

Clearly, the rates of reactions 2 and 3 are equal, and the rate of formation of HBr is twice the rate of reaction 2, so we may write

$$\frac{d[HBr]}{dt} = 2k_2[Br][H_2] = 2k_2k_1^{1/2}[H_2][Br_2]^{1/2} = k[H_2][Br_2]^{1/2}$$

As the reaction proceeds, and HBr is present in appreciable quantities, reverse reactions 2 and 3 occur, and a more complicated rate law can be derived. Remarkably, this law had previously been found experimentally by Bodenstein and Lind (13).

It has been found that atoms and free radicals (such as OH, HCO, CH_3, and C_2H_5) are important intermediates in many reactions. Frequently their concentrations rapidly reach steady states which can be calculated as above for H and Br atoms, and these steady values change slowly as the reactants are used up. Reactions 2 and 3 of the above series form a reaction chain, in which the product of one reaction (H, for example) is a reactant in the next, and vice versa. That is, even if reaction 1 suddenly stopped, reactions 2 and 3 would continue to produce HBr by this cyclic, chain process. This type of *chain reaction*, in which the concentration of the reactive species becomes steady, is often called a *linear chain*. In a number of combustion reactions branching chains are encountered, propagated by reaction such as

$$O + H_2 \rightarrow OH + H$$

in which two active species are produced for each one used up. The growth in the rate of reaction via this type of process leads to an explosion. Chain reactions can sometimes be started photochemically, the light energy dissociating a reactant molecule to form atoms or free radicals.

Let us look at a few gaseous free radical reactions in some detail. For the overall reaction

$$2\ CH_4 \rightarrow C_2H_6 + H_2$$

the reaction mechanism probably consists of the following three elementary steps

$$CH_4 \xrightarrow{1} CH_3 + H$$

$$H + CH_4 \xrightarrow{2} CH_3 + H_2$$

$$2\ CH_3 \xrightarrow{3} C_2H_6$$

Let us, for simplicity, overlook the experimental fact that ethane is unstable, and would react further. If we assume that only the above three reactions occur, then let us make the steady-state approximation and calculate the concentrations of CH_3 and H:

$$\frac{d[CH_3]}{dt} = k_1[CH_4] + k_2[H][CH_4] - 2k_3[CH_3]^2 = 0$$

$$\frac{d[H]}{dt} = k_1[CH_4] - k_2[H][CH_4] = 0$$

From the second equation, we obtain

$$[H] = \frac{k_1}{k_2}$$

and if this relationship is substituted into the first steady-state equation we obtain

$$[CH_3] = \left(\frac{k_1}{k_3}[CH_4]\right)^{1/2}$$

These expressions for [H] and $[CH_3]$ can be used to obtain the rates of all of the subreactions in terms of $[CH_4]$. We have

$$\text{rate (1)} = k_1[CH_4]$$

$$\text{rate (2)} = k_2 \cdot \frac{k_1}{k_2}[CH_4] = k_1[CH_4]$$

$$\text{rate (3)} = k_3 \frac{k_1}{k_3}[CH_4] = k_1[CH_4]$$

That is, the rates of all of the subreactions turn out to be equal, and equal to the rate of the overall reaction for the formation of ethane. (This equality is not, of course, typical of all free radical mechanisms.) It is interesting to note that the steady-state concentration of H atoms is independent of the concentration of methane, so that the ratio of [H] to $[CH_4]$ increases linearly as $[CH_4]$ decreases. A smaller increase of $[CH_3]$ relative to $[CH_4]$ also occurs as $[CH_4]$ decreases. This effect is typical of free-radical reactions. That is, since combination of radicals is a second- or third-order process, while dissociation is a first- or second-order process, dissociation is favored relative to combination at low pressures. If the pressure is lowered sufficiently, there would finally be too little methane to produce the amount required by the steady state, so that the steady state would never be reached. Rather, [H] would rise to some maximum value smaller than k_1/k_2, then decrease to an equilibrium value.

Let us consider the reverse reaction and the equilibrium condition. If we start with equimolar amounts of C_2H_6 and H_2, and no CH_4, we can write two different steady-state equations:

$$\frac{d[CH_3]}{dt} = 2k_{-3}[C_2H_6] - k_{-2}[CH_3][H_2] - k_{-1}[CH_3][H] = 0$$

$$\frac{d[H]}{dt} = k_{-2}[CH_3][H_2] - k_{-1}[CH_3][H] = 0$$

From the second equation we obtain

$$[H]_r = \frac{k_{-2}}{k_{-1}}[H_2]$$

while substitution of this expression for [H] into the first equation gives

$$[CH_3]_r = \frac{k_{-3}}{k_{-2}}\frac{[C_2H_6]}{[H_2]} = \frac{k_{-3}}{k_{-2}}$$

under our conditions for which $[C_2H_6]$ and $[H_2]$ remain equal (that is, we assume at this point that the concentrations of H and CH_3 are so small that they do not noticeably upset the stoichiometry of the overall reaction). As before, rates of all three subreactions and the overall reaction are equal, the value being

$$\text{rate}_r = k_{-3}[C_2H_6]$$

Both the steady-state concentrations and the reverse rate are different from those for the forward reaction. Moreover, the two calculated rates would not be equal at equilibrium.

We may also calculate [H], [CH_3], and the rates of each subreaction under equilibrium conditions if we again assume that [H] and [CH_3] are small enough not to disturb the overall stoichiometry. We obtain

$$[\mathrm{H}]_{\mathrm{eq}} = \frac{k_1 k_3^{1/2}[\mathrm{CH_4}]}{k_{-1}k_{-3}^{1/2}[\mathrm{C_2H_6}]^{1/2}}$$

and

$$[\mathrm{CH_3}]_{\mathrm{eq}} = \left(\frac{k_{-3}[\mathrm{C_2H_6}]}{k_3}\right)^{1/2}$$

while

$$\text{rate (1)} = \text{rate }(-1) = k_1[\mathrm{CH_4}]$$

$$\text{rate (2)} = \text{rate }(-2) = \frac{k_{-2}k_{-3}^{1/2}[\mathrm{C_2H_6}]^{3/2}}{k_3^{3/2}}$$

$$\text{rate (3)} = \text{rate }(-3) = k_{-3}[\mathrm{C_2H_6}]$$

The equilibrium concentrations are different from either of the steady-state values; while for each reaction, although forward and reverse rates are equal, the numerical values of the rates differ from one reaction to another.

It is clear that the calculation of steady-state concentrations of intermediates based on initial concentrations is open to question on several counts:

(a) How long does it take, at the beginning of the reaction, for the steady-state concentrations to build up? Is there an appreciable time at the beginning during which the reaction is proceeding slowly since the concentrations have not yet reached their steady values?

(b) Does the reaction approach close enough to equilibrium so that the steady-state concentrations begin to change due to the influence of the reverse reactions?

(c) Are the concentrations of intermediates low enough that they do not perturb the overall stoichiometry to a noticeable amount?

All of these points should be checked as well as possible in a kinetic study in which free radicals play a role. In fact, there are many situations in which they can be answered in a favorable sense—that is, very small steady-state concentrations of radicals are set up very quickly, and there is a substantial period of reaction during which the overall rate is controlled by the forward subreactions only.

If one wishes to obtain a complete picture of the progress of the reaction, including the build-up of the free radicals and the approach to equilibrium, it is necessary to include both forward and reverse rate constants of all three reactions in the calculation. No simple expressions of the type we have derived for special situations now arise, and the best approach is to carry out a numerical integration over the time of interest. A computer program to do this is described in Fig. 4.4 and Table 4.3, while the results of the calculation for a typical high-temperature experiment are shown in Fig. 4.5.

The pyrolysis of ethane is also a free-radical process, but the mechanism turns out to be different in many respects from that of methane. There are many more steps, of which the most important are:

$$C_2H_6 \xrightarrow{1} 2\,CH_3$$

$$CH_3 + C_2H_6 \xrightarrow{2} CH_4 + C_2H_5$$

$$C_2H_5 \xrightarrow{3} C_2H_4 + H$$

$$H + C_2H_6 \xrightarrow{4} C_2H_5 + H_2$$

$$C_2H_5 + C_2H_5 \xrightarrow{5} C_4H_{10} \quad \text{or} \quad C_2H_4 + C_2H_6$$

$$H + C_2H_5 \xrightarrow{6} C_2H_6$$

The pyrolysis has been studied most extensively at temperatures of 800–900°K, in static experiments lasting the order of an hour (13). Products of the reaction are primarily C_2H_4 and H_2, while small amounts of CH_4 and larger hydrocarbons (C_3H_8 and C_4H_{10}) are found. The predominance of C_2H_4 and H_2 indicate that reactions 3 and 4 occur more rapidly than do the others.

We should pause briefly to note some of the terminology commonly used to describe such chain reactions. Reactions 3 and 4, which produce most of the reaction, are called *chain-propagating* steps. As a pair they are self-perpetuating since each one produces a new radical for each one that is used up. Reaction 1 is a *chain-initiating* step, while reactions 5 and 6 are *chain-terminating*, for obvious reasons. It is not uncommon to have two possible reactions between two radicals, as indicated for reaction 5. *Combination* of the radicals, in this instance to form C_4H_{10}, typically occurs with very low (often negligibly small) activation energy, so is favored at low temperatures. Transfer of an atom, or *disproportionation*, typically does

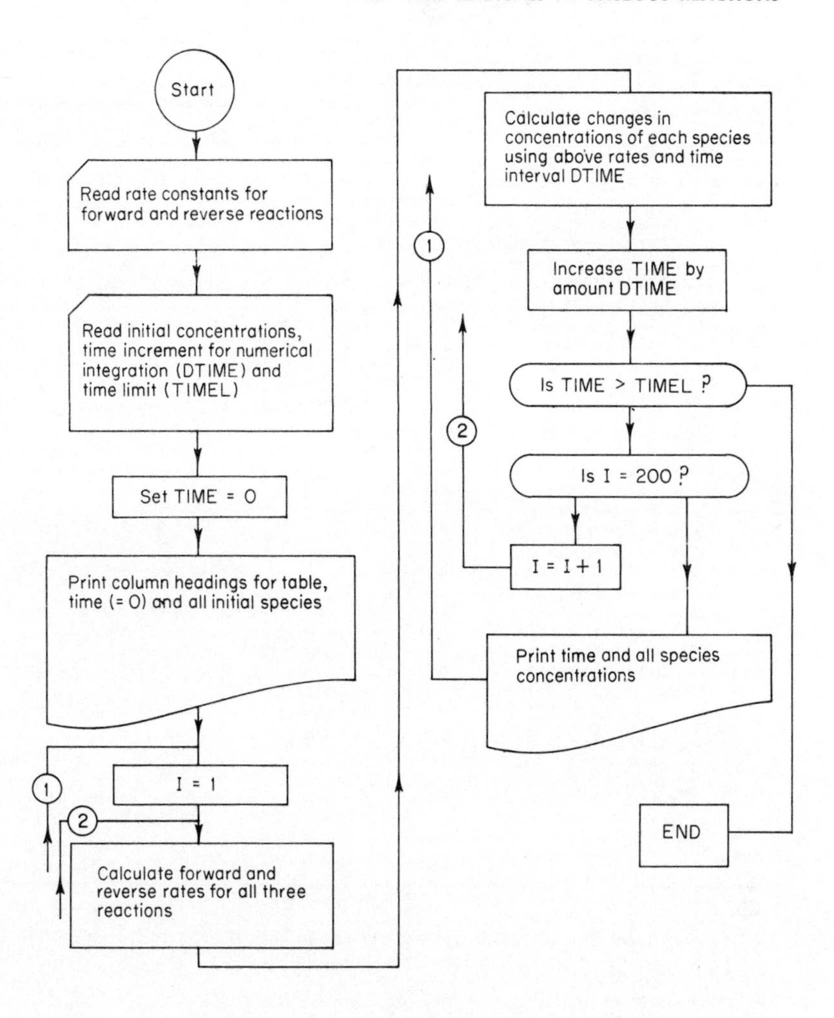

Fig. 4.4 Logic diagram for a FORTRAN program for calculating species concentrations in the first stages of methane decomposition.

involve a small activation energy, so this process is not favored at low temperatures, but becomes more likely as the temperature is raised. At low pressures and high temperatures, we would expect unimolecular reactions like 1 to become second order, while reactions like 6, which are the reverse of unimolecular reactions, may become third order. Under given conditions, second-order kinetics are more likely for unimolecular reactions of smaller

Table 4.3 A FORTRAN Program to Calculate Product Yields in the First Stage of Methane Pyrolysis

```
C         PROGRAM METH
C         REACTIONS ARE 1          CH4 = CH3 + H
C                       2      H + CH4 = CH3 + H2
C                       3         2CH3 = C2H6
C         FIRST READ RATE CONSTANTS OF FORWARD AND REVERSE REACTIONS
          READ(5,100)RK1,RR1,RK2,RR2,RK3,RR3
   100 FORMAT(6E12.4)
C         READ CONCENTRATIONS, TIME INCREMENT FOR NUMERICAL INTEGRATION
C         AND TOTAL TIME
          READ(5,101)CH4,CH3,H,C2H6,H2,DTIME,TIMEL
   101 FORMAT(7E10.4)
          TIME = 0.
          WRITE(6,200)
   200 FORMAT(1H1,'           TIME            CH4              CH3              H
      1      C2H6            H2  ')
          WRITE(6,201)TIME,CH4,CH3,H,C2H6,H2
   201 FORMAT(1X,6E13.4)
    12 DO 10 I=1,200
          F1 = RK1*CH4
          B1 = RR1*CH3*H
          F2 = RK2*H*CH4
          B2 = RR2*CH3*H2
          F3 = RK3*CH3*CH3
          B3 = RR3*C2H6
C
          CH4 = CH4 + DTIME*(-F1+B1-F2+B2)
          CH3 = CH3 + DTIME*(F1-B1+F2-B2-2.*F3+2.*B3)
          H = H + DTIME*(F1-B1-F2+B2)
          C2H6 = C2H6 + DTIME*(F3-B3)
          H2 = H2 + DTIME*(F2-B2)
          TIME = TIME + DTIME
          IF(TIME - TIMEL)10,10,11
    10 CONTINUE
          WRITE(6,201)TIME,CH4,CH3,H,C2H6,H2
          GO TO 12
    11 STOP
          END
```

rather than larger molecules. As an example, under conditions in which reaction 6 is third order, reaction 5 could still be second order.

The experimental results at 800–900°K and pressures of 0.1–1 atm can be explained fairly well by forward reactions 1–5, assuming steady states of radical concentrations. This assumption leads to the following equations:

$$\frac{d[CH_3]}{dt} = 2k_1[C_2H_6] - k_2[C_2H_6][CH_3] = 0$$

$$\frac{d[H]}{dt} = k_3[C_2H_5] - k_4[H][C_2H_6] = 0$$

$$\frac{d[C_2H_5]}{dt} = k_2[C_2H_6][CH_3] - k_3[C_2H_5] + k_4[H][C_2H_6] - 2k_5[C_2H_5]^2 = 0$$

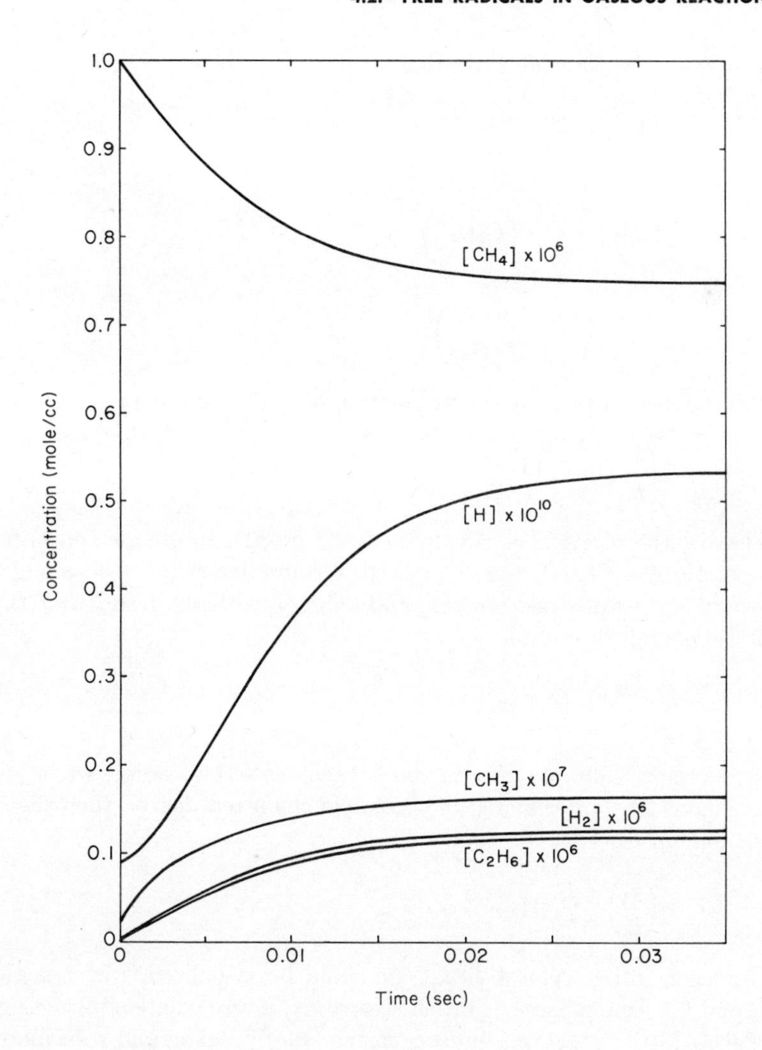

Fig. 4.5 Species concentrations in the pyrolysis of methane, calculated by the computer program of Table 4.3, using the following rate constants, which are applicable at about 1600°K at high pressure:

RK1 = 14 sec^{-1}

RR1 = 1.2×10^{13} $mole^{-1}$ cc sec^{-1}

RK2 = 1.5×10^{12} $mole^{-1}$ cc sec^{-1}

RR2 = 2.9×10^{10} $mole^{-1}$ cc sec^{-1}

RK3 = 2.0×10^{13} $mole^{-1}$ cc sec^{-1}

RR3 = 4.5×10^{4} sec^{-1}

from which we can calculate that

$$[CH_3] = \frac{2k_1}{k_2}$$

$$[C_2H_5] = \left(\frac{k_1}{k_5}[C_2H_6]\right)^{1/2}$$

$$[H] = \left(\frac{k_3}{k_4}\frac{k_1}{k_5[C_2H_6]}\right)^{1/2}$$

and the overall rate for the production of C_2H_4 (or H_2) is

$$k_3\left(\frac{k_1}{k_5}[C_2H_6]\right)^{1/2}$$

That is, the reaction is calculated to be $\frac{1}{2}$ order in ethane concentration. If, on the other hand, reaction 6 is the terminating step, we obtain different steady-state equations for $[H]$ and $[C_2H_5]$, with the result that the rate of the overall reaction is

$$\left(\frac{k_1k_3k_4}{k_6}\right)^{1/2}[C_2H_6]$$

In one more example, if reaction 1 becomes second order, while reaction 5 remains first order and is the principal chain terminater, then the rate of the overall reaction is

$$k_3\left(\frac{k_1}{k_5}\right)^{1/2}[C_2H_6]$$

Numerous other typical situations could be visualized, with reactions 3, 5, and 6 becoming second order. Moreover, intermediate situations can be expected, with reactions midway in the fall-off region, and with more than one terminating reaction being important. Accordingly, we would expect that under many conditions, the reaction order would gradually change with temperature and pressure.

The overall activation energy can be found from the individual k's involved. For example, for the situation where the rate is

$$k_3\left(\frac{k_1}{k_5}\right)^{1/2}[C_2H_6]$$

the overall activation energy will be $E_3 + \frac{1}{2}E_1 - \frac{1}{2}E_5$, and will take on corresponding values for the other rate laws. Accordingly, the activation

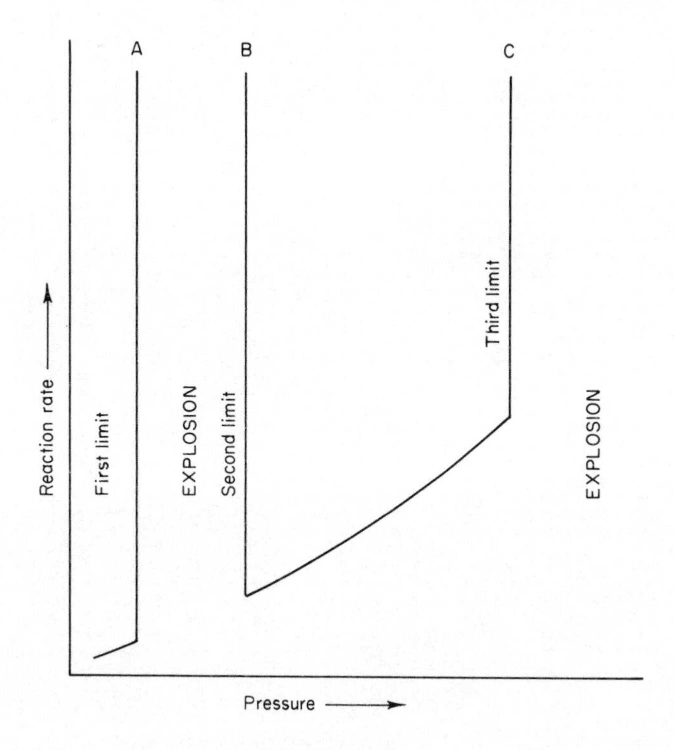

Fig. 4.6 The relationship between the rate of the H_2–O_2 reaction and the pressure in a particular reaction vessel at fixed composition and temperature (not to scale) (15).

energy will tend to change with the rate law. As with the methane example, one is driven to a computer solution if he wishes to correlate data over a wide range of temperatures, pressures, and concentrations. At the same time, the "special case" calculations are a valuable guide as to what one may expect.

Let us look at one more free-radical gas reaction, hydrogen oxidation. In the first kinetic studies, carried out by C. N. Hinshelwood and his students during the second quarter of this century, a characteristic pattern emerged from experiments carried out in static bulb reactors in the 750–850°K range (15–17). The rate of reaction appeared to depend on pressure in a discontinuous way, as indicated in Fig. 4.6. At low pressures the rate of reaction was slow enough to be measured in minutes or hours (depending on the temperature), but as the pressure rose to some characteristic value, typically a few Torr, an explosion suddenly occurred. An attempt to fill the bulb with reaction mixture to a pressure above this *first explosion limit*

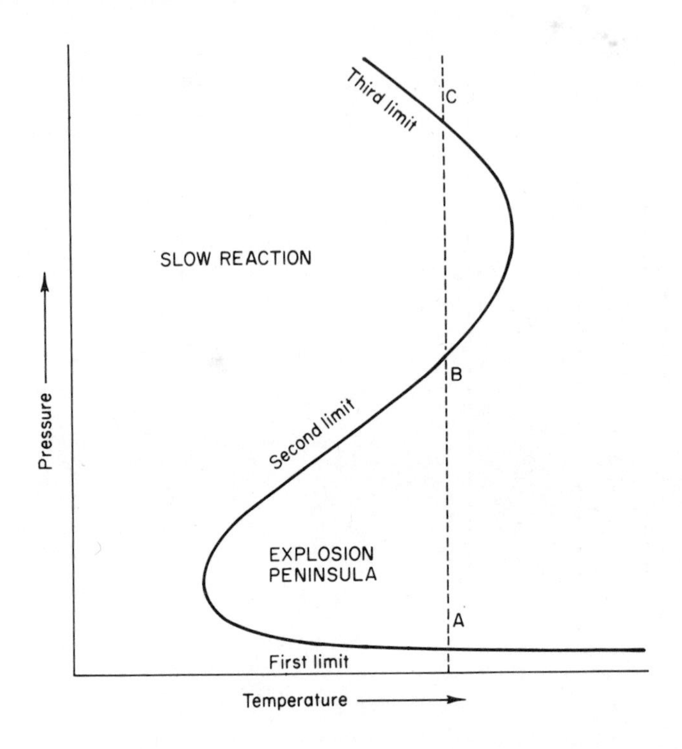

Fig. 4.7 The effect of temperature on the explosion limits of a particular hydrogen–oxygen mixture in a particular reaction vessel (not to scale) (15).

led to apparently instantaneous reaction as soon as the pressure reached the limit. The first explosion limit pressure was a function of stoichiometry, temperature, and also of the size, shape, and material of the container, indicating that heterogeneous reactions are important.

As pressures were increased still further (experimentally this could be done by pressurizing the mixture at low temperature, than heating it), there was a *second explosion limit* at which the rate of reaction suddenly became much slower, so that it could be measured by conventional means. As indicated in Fig. 4.6, the new rate was normally higher than that found at low pressures. The rate continued to increase as the pressure was raised, in some instances in such a way that the temperature would rise rapidly due to the heat of reaction and a *thermal explosion* would occur, in which the rise in reaction rate due to rise in temperature would lead to an increasing and unmeasurably fast rate of reaction. However, they showed that in some instances a distinct *third explosion limit* could be detected; this limit was not surface dependent, and at it, with increasing pressure, a sudden increase in the reaction rate to an unmeasurable value occurred.

A set of such diagrams at several temperatures led to Fig. 4.7 which shows (qualitatively) the boundary, as a function of both temperature and pressure, between the region of moderately slow reaction and apparently instantaneous explosion for a particular stoichiometry in a particular reaction vessel.

Since then, some of the "unmeasurably fast" reaction rates reported by Hinshelwood have been studied by shock tube techniques, at temperatures as high as 3500°K. Typically, reaction times of 10 μsec to 10 millisec have been measured. Reaction rates are not constant during those periods, but increase exponentially with time following a short initiation period. Unlike Hinshelwood's data, the shock tube results are entirely homogeneous, and it is probably most understandable to look at the explanation of them first, before the more complex bulb experiments.

The most important reactions in the H_2–O_2 system are the following:

$$H_2 + O_2 \xrightarrow{1} 2\,OH$$

$$OH + H_2 \xrightarrow{2} H_2O + H$$

$$H + O_2 \xrightarrow{3} OH + O$$

$$O + H_2 \xrightarrow{4} OH + H$$

$$H + O_2 + M \xrightarrow{5} HO_2 + M$$

$$HO_2 + H_2 \xrightarrow{6} H_2O_2 + H$$

$$H_2O_2 + M \xrightarrow{7} 2\,OH + M$$

$$H + OH + M \xrightarrow{8} H_2O + M$$

$$H + H + M \xrightarrow{9} H_2 + M$$

$$O + O + M \xrightarrow{10} O_2 + M$$

In these reactions, M represents any molecule present that can act as an energy transfer agent. That is, under most of the conditions studied the unimolecular decompositions of HO_2, H_2O_2, H_2O, H_2, and O_2 will be in the second-order region.

At the high-temperature end of the shock tube range—say above 1500°K—the reaction mechanism is particularly simple. To begin, a small number of OH radicals is produced by reaction 1, the OH concentration temporarily leveling off when the rate of production by reaction 1 equals the rate of loss by reaction 2. The initial steady-state OH concentration is thus given by the relationship

$$2k_1[H_2][O_2] - k_2[OH]_0[H_2] = 0$$

so that

$$[OH]_0 = \frac{2k_1[O_2]}{k_2}$$

From this point on, reactions 2, 3, and 4 take over and bring about an exponential rate of reaction, during which [H], [OH], and [O] maintain a constant ratio to one another. That is, it is approximately true that at a given time the rate of production of O atoms by reaction 3 is equalled by the rate of loss by reaction 4, so that

$$k_3[H][O_2] = k_4[O][H_2]$$

and

$$\frac{[O]}{[H]} = \frac{k_3[O_2]}{k_4[H_2]}$$

Similarly, it is approximately true that for OH,

$$k_3[H][O_2] + k_4[O][H_2] = k_2[OH][H_2]$$

$$2k_3[H][O_2] = k_2[OH][H_2]$$

and

$$\frac{[OH]}{[H]} = \frac{2k_3[O_2]}{k_2[H_2]}$$

Of course, we cannot solve for absolute steady-state values since none exist. However, we can conclude, from the above statements, that the total free-radical concentration is $C[H]$, where

$$C = 1 + \frac{k_3[O_2]}{k_4[H_2]} + \frac{2k_3[O_2]}{k_2[H_2]}$$

and that the rate of growth of the total concentration of free radicals is $2k_3[H][O_2]$ since both reactions 3 and 4, which have the same rate, contribute to the increase in the total number of radicals. Therefore

$$\frac{d(C[H])}{dt} = 2k_3[H][O_2]$$

$$\frac{d[H]}{[H]} = \frac{2k_3[O_2]}{C}\,dt$$

If we assume there is some initial value of [H] at time zero, approximately given by

$$[H]_0 = \frac{k_1[H_2]}{k_3}$$

then we can write, after integration,

$$[H] = [H]_0 \exp(2k_3[O_2]t/C)$$

Similar exponential curves exist for [OH] and [O] (since in this part of the reaction the ratios of the radical concentrations remain constant).

At first, the exponential growth of the reaction rate is undetectable with laboratory instruments, but after an apparent "induction period" the reaction rather suddenly (since most instruments have a linear response) becomes noticeable, as heat and light are produced, and the free radicals can be observed by absorption spectroscopy (see Fig. 4.8). The end of the exponential growth period comes as the concentration of H_2 and/or O_2 is significantly depleted by reaction: in this last stage the rate of reaction reaches a maximum, then diminishes as equilibrium is approached via combination reactions such as 8, 9, and 10. Since reaction 3 has an activation energy of 17 kcal, for a given reaction mixture the lengths of the induc-

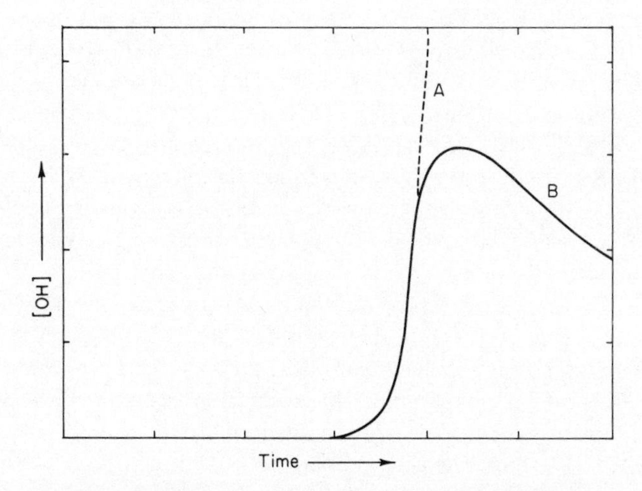

Fig. 4.8 The exponential growth of [OH] in the hydrogen–oxygen reaction, as observed by a linear-response instrument. Curve A: continuation of exponential curve assuming no reduction in H_2 and O_2 concentrations. Curve B: more realistic curve reflecting peaking and fall-off of [OH] due to depletion of H_2 and O_2 and combination of OH and H to form water vapor.

tion periods also appear to have an "activation energy" of about this amount; the times, of course, decrease with increasing temperature.

At lower temperatures, reactions 5, 6, and 7 become important. The effect of reaction 5 is to convert reactive H into unreactive HO_2, in competition with reaction 3, which converts H into two free radicals. In the absence of reactions 6 and 7, reaction 5 would stop chain branching altogether if it removed radicals as fast as they were being multiplied, that is, if

$$k_5[H][O_2][M] = 2k_3[H][O_2]$$

or if

$$k_5[M] = 2k_3$$

This equality will undoubtedly occur at some temperature since the temperature dependence of k_5 is small, and therefore graphs of $k_5[M]$ and $2k_3$ versus temperature are bound to cross (at 1000°K for pressures of the order of 1 atm). Accordingly, as the temperature is lowered, experimental induction times rise above the extrapolation of the high-temperature curve, but since reactions 6 and 7 do cause HO_2 radicals to be converted slowly back into reactive species, the effect is less abrupt than would be predicted by reaction 5 alone.

The observations of Hinshelwood can be explained in terms of the above reactions, coupled with the idea that the walls of his containers catalyzed some reactions that would not occur significantly in the gas phase. Lewis and Von Elbe (17) consider that the most important heterogeneous initiating reaction is the formation of H_2O_2, which they consider occurs at a rate little affected by the concentrations of H_2 and O_2 in the gas phase (this is not unreasonable, as we will see later, for heterogeneous reactions). The H_2O_2 will be desorbed, and either decompose homogeneously by reaction 7, or return to the surface and be catalytically decomposed. The rate of formation of gaseous OH radicals will depend on the total gas pressure, and at low pressures will be directly proportional to the pressure.

Under Hinshelwood's relatively low-temperature conditions, H atoms will be much more stable than OH and O, so that when the last two are formed they are transformed rapidly, via reactions 2 and 4, into H atoms and water molecules. This comes about since the activation energies of reactions 2 and 4 are 5 and 9 kcal respectively, compared to 17 kcal for reaction 3. The chance of H atoms diffusing to the wall are much greater than the chances of O and OH doing so. Accordingly, the fairly immediate result of the decomposition of an H_2O_2 molecule

$$H_2O_2 + M \rightarrow 2\,OH + M$$

is the formation of two H and two H_2O molecules by reaction 2

$$2\,OH + 2\,H_2 \rightarrow 2\,H_2O + 2\,H$$

so we could say that the rate of production of H atoms by wall catalysis is

$$\frac{d[\mathrm{H}]}{dt} = 2k_7[\mathrm{H_2O_2}][\mathrm{M}]$$

where $[H_2O_2]$ will be fairly constant at low pressures.

Occasionally, and to an increasing extent as the pressure increases, H atoms will react by reaction 3. This reaction will be followed rapidly by reaction of the O atom by reaction 4, and of the two OH radicals by reaction 2, so the net result will be

$$\mathrm{H + O_2 + 3\ H_2 \rightarrow 2\ H_2O + 3\ H}$$

Of course the rate of this nonelementary reaction will be governed by that of reaction 3.

Other reactions involving H atoms, occurring at higher pressures, are reaction 5, which forms HO_2, and also reaction 6, which will tend to be followed rapidly by reactions 7 and 2 to give

$$\mathrm{HO_2 + 3\ H_2 \rightarrow 2\ H_2O + 3\ H}$$

Finally, when H atoms diffuse to the container walls it seems that they are converted to stable molecules (H_2O or H_2) by a process which is first order in [H], while similar conversion of HO_2 occurs when it reaches the wall.

We can now write a reasonably complete equation for the rate of change of [H] with time:

$$\frac{d[\mathrm{H}]}{dt} = 2k_7[\mathrm{H_2O_2}][\mathrm{M}] - k_\mathrm{w}[\mathrm{H}] + 2k_3[\mathrm{H}][\mathrm{O_2}]$$

$$- k_5[\mathrm{H}][\mathrm{O_2}][\mathrm{M}] + 3k_6[\mathrm{HO_2}][\mathrm{H_2}] + 2k_1[\mathrm{H_2}][\mathrm{O_2}]$$

where k_w is the rate constant for catalytic removal of H atoms at the walls. In this equation we have implicitly assumed that the gas composition is uniform—otherwise we could not quote species concentrations. Whether or not a steady-state concentration of H atoms can be achieved depends on the relative magnitudes of the terms in the above equation.

At low pressures, the first two terms dominate, and a steady-state concentration of H atoms

$$[\mathrm{H}] = \frac{2k_7[\mathrm{H_2O_2}][\mathrm{M}]}{k_\mathrm{w}}$$

is found. The approximately constant $[H_2O_2]$ leads to [H] being approximately proportional to the total pressure, so the reaction shows first-order behavior. This is the behavior shown in the first part of Fig. 4.6, at pressures below 1 Torr.

A moderate increase in pressure leads to noticeable contribution by reaction 3. If it is included in the steady-state equation, we obtain

$$[H] = \frac{2k_7[H_2O_2][M]}{k_w - 2k_3[O_2]}$$

At first, then, by decreasing the denominator, reaction 3 simply increases the rate of the steady reaction, but as the pressure increases, eventually

$$2k_3[O_2] = k_w$$

and at that point a steady state [H] is no longer possible, but [H] grows exponentially, leading to an explosion in less than a second. This transition from a steady state to exponential growth, therefore, causes the first explosion limit of Fig. 4.6. Clearly, it is highly dependent on the container geometry and material, which determine k_w.

At higher pressures the rate of reaction 5, which depends on the second power of the pressure, may become large compared to that of reaction 3. If we include reaction 5 in the equation for [H], we find

$$[H] = \frac{2k_7[H_2O][M]}{k_w - 2k_3[O_2] + k_5[O_2][M]}$$

Clearly, if $k_w + k_5[O_2][M] > 2k_3[O_2]$, then [H] can become finite and stable again. Usually, at the higher pressures (100 Torr) at which this happens, k_w has become negligible in comparison to the other two reactions, so that usually the condition for the second explosion limit is given as

$$k_5[M] = 2k_3$$

a relationship given before in connection with shock tube data. This limit, then, is not strongly dependent on the apparatus. At pressures above the second explosion limit, steady reaction is maintained because the HO_2 radicals produced by reaction 5 are stable enough to diffuse to the walls.

At still higher pressures, the number of collisions undergone by the HO_2 becomes sufficient that a substantial fraction of them react to reform H atoms by reaction 6. In effect, this cuts down on the effectiveness of reaction 5, and at a sufficiently high pressure a second transition from steady [H] to exponential growth causes the third explosion limit.

Detailed kinetic analysis of such experiments is very difficult partly because the rate laws for the surface reactions tend to be complex and to change with pressure (see Chapter 6) and partly because concentration and temperature gradients that are hard to measure and to treat mathematically can develop in static reactors. Accordingly, while the qualitative explanations of the explosion limits given above were deduced by Hinshelwood, actual values of the rate constants of the elementary reactions were not obtained until the shock tube experiments were carried out under homogeneous conditions, within the past 15 years.

References

1. F. W. Schneider and B. S. Rabinovitch, *J. Amer. Chem. Soc.* **84,** 4215 (1962).
2. F. A. Lindemann, *Trans. Faraday Soc.* **17,** 598 (1922).
3. O. K. Rice and H. C. Ramsperger, *J. Amer. Chem. Soc.* **49,** 1617 (1927); **50,** 617 (1928).
4. L. S. Kassel, *J. Phys. Chem.* **32,** 225 (1928).
5. R. A. Marcus, *J. Chem. Phys.* **20,** 359 (1952).
6. E. V. Waage and B. S. Rabinovitch, *J. Chem. Phys.* **52,** 5581 (1970).
7. E. V. Waage and B. S. Rabinovitch, *Chem. Rev.* **70,** 377 (1970).
8. T. Bayer and D. F. Swinehart, *Commun. ACM* **16,** 379 (1973).
9. S. E. Stein and B. S. Rabinovitch, *J. Chem. Phys.* **58,** 2438 (1973).
10. J. A. Christiansen, *Kgl. Dan. Vidensk. Selsk. Mat. Fys. Medd.* **1,** 14 (1919).
11. K. F. Herzfeld, *Z. Elektrochem. Angew. Phys. Chem.* **25,** 301 (1919); *Ann. Physik. (Leipzig)* **59,** 635 (1919).
12. M. Polanyi, *Z. Elektrochem. Angew. Phys. Chem.* **26,** 50 (1920).
13. M. Bodenstein and S. C. Lind, *Z. Physik. Chem. Stoechiom. Verwandschattslehre* **57,** 168 (1907).
14. M. C. Lin and M. H. Back, *Can. J. Chem.* **44,** 505 (1966).
15. C. N. Hinshelwood and A. T. Williamson, "The Reaction between Hydrogen and Oxygen." Oxford Univ. Press, London and New York, 1934.
16. J. A. Barnard, *Sci. Progr. (London)* **47,** 30 (1959).
17. B. Lewis and G. von Elbe, "Combustion, Flames and Explosions of Gases." 2d Ed., Academic Press, New York, 1961.

Further Reading

D. L. Bunker, "Theory of Elementary Gas Reaction Rates." Pergamon Press, New York, 1966 (good short presentation).

H. S. Johnston, "Gas Phase Reaction Rate Theory." Ronald Press, New York, 1966 (more detailed and advanced presentation).

L. S. Kassel, "Kinetics of Homogeneous Gas Reactions." Reinhold, New York, 1932 (early presentation by the K of the RRKM theory).

E. E. Nikitin, "Theory of Thermally Induced Gas Phase Reactions." Indiana Univ. Press, Bloomington, Indiana, 1965 (translation of a Russian text).

P. J. Robinson and K. A. Holbrook, "Unimolecular Reactions." Wiley (Interscience), New York, 1972 (an up-to-date review of theory and experiment in this area).

Problems

4.1 For the unimolecular reaction of methyl isocyanide to acetonitrile ($CH_3NC \rightarrow CH_3CN$) referred to in the text, Schneider and Rabinovitch found an Arrhenius equation for the rate constant at high pressure of

$$k_\infty = 4.0 \times 10^{13}\, e^{-38,400/RT}$$

Follow through the discussion of the text to obtain the simple Lindemann fall-off curve at 230.4°C. Obtain E_{CT} and k_2 from their equation, calculate k_1 and k_{-1} via the collision theory using their value of $\sigma = 4.5$ Å, and obtain a series of apparent first-order rate constants, which by Eq. (4.2) would be

$$k_{app} = \frac{k_1 k_2[A]}{k_{-1}[A] + k_2}$$

where A stands for methyl isocyanide, and concentrations are in mole/cc. Plot the points on a graph and compare your results with Fig. 4.1.

4.2 With reference to the example of Table 4.1,

(a) What five quantum states of the supposed molecule have energies of 1300 cm^{-1}?

(b) What would be the multiplicity of states at 3000 cm^{-1} energy?

(c) What is $G(E)$ at 2000 cm^{-1}?

(d) What is Q_{PD} (assuming all three degrees of freedom can contribute to reaction) at 500°K?

(e) What is $B(E)$ at 1500 cm^{-1}, at 500°K? (Note that your answer here could depend on your method of averaging.)

4.3 Show how Eq. (4.21) is obtained from Eq. (2.41).

4.4 Show, by reference to more familiar derivatives, that

$$\frac{d \ln T}{d(1/T)} = -T$$

4.5 Show in detail how Eq. (4.23) follows from Eq. (3.54).

4.6 For the reaction

$$NO_2 \rightarrow NO + O$$

the rate constant has been measured as a function of pressure by J. Troe (*Ber. Bunsenges. Phys. Chem.* **73,** 144 (1969)). The data were obtained in a shock tube at high temperatures and pressures, the NO_2 being highly diluted with argon for experimental reasons, so that the bulk of intermolecular collisions were between Ar and NO_2, rather than NO_2–NO_2. From a smooth curve drawn through the data at 1540°K, the following values are found:

log[Ar] (mole cc^{-1})	log k_1, apparent (sec^{-1})
−5.0	1.96
−4.0	2.96
−3.5	3.46
−3.0	3.80
−2.5	3.99

He estimates that log k_∞ is about 4.10 at this temperature.

If we wish to interpret these data by the RRKM theory, we need certain molecular data. NO_2 is a bent molecule, so it has three different moments of inertia, which are 3.5, 63.6, and 67.1 $\times$ 10^{-40} g cm^2. The small moment corresponds to rotation about an axis parallel to a line joining the O atoms; it would go to zero if the molecule were straightened out. There are three vibrational frequencies, of 1322, 750, and 1616 cm^{-1}. These are analogous to those of the linear triatomic molecule of Fig. 3.4; the 750-cm^{-1} bending vibration would become doubly degenerate if the molecule were straightened. For simplicity in finding $N(E)$, let us round these frequencies off to 1300, 800, and 1600 cm^{-1}.

The zero-point energy of dissociation for the reaction, ΔE_0, is 72.78 kcal, known quite accurately. For such an endothermic reaction, we would expect that E_0, the zero-point activation energy, would be close to ΔE_0. To make the energy compatible with the vibrational frequencies, let us write this as 25,500 cm^{-1}.

(a) Start by making an ACT calculation to check k_∞. Since the activated complex has the same molecular weight as the molecule, translational contributions to the partition functions will cancel out. By one of the methods available in the literature it may be estimated that the large moments of inertia increase to about 90 $\times$ 10^{-40} g cm^2 each due to lengthening of one N—O bond, while if the O—N—O angle increased slightly, the small moment might remain the same. This simple approach gives the ratio of rotational partition functions as $90/(64 \times 67)^{1/2}$, with the 90 $\times$ 10^{-40} g cm^2 value being somewhat adjustable, if necessary. If the data cannot be interpreted this way, then a more radical approach, such as assuming a linear complex, might be tried. The remaining parts of the partition functions are the vibrational contributions, which can easily be obtained for the molecule (use the rounded frequencies). For the complex the third frequency would be the reaction coordinate, while the other two would be lower than those in the molecule, due to weakening of the N—O bond. Since either oxygen atom can be lost, the statistical factor is 2. Complete the ACT calculation, using two complex frequencies that give the correct k_∞ at 1540°K.

(b) Obtain $N(E)$ values for the molecule at 26,000–40,000 cm^{-1} at every 1000 cm^{-1}, and calculate $B(E)$ values at 1540°K.

(c) Obtain $G(E)$ values for the complex at 500–14,500 cm^{-1}, at every 1000 cm^{-1} (corresponding to the $N(E)$ values of part b), and calculate k_E values at these energies.

(d) By numerical integration obtain k_∞ using Eq. (4.16). This should be equal (allowing for arithmetic round-off errors and some distortion due to the rather large step size of 1000 cm^{-1}) to the result of part a.

(e) Calculate k_{-1} via Eq. (4.6) (with $E_{CT} = 0$) and using a collision diameter σ of 4 Å.

(f) By numerical integration using Eq. (4.11), calculate k_1 at each of the experimental concentrations, and compare your calculations with experiment. If you find your fall-off occurs at lower pressures than found experimentally, the answer may be (as has been found for other systems) that argon is less efficient than NO_2 as energy transfer agent and that, for example, 10^{-3} mole cc^{-1} of argon may be equivalent to 10^{-4} mole cc^{-1} of NO_2. Perhaps your calculated curve can match the experimental one by a translation along the concentration axis. If so, the amount of translation will be a measure of the Ar efficiency.

(g) Use appropriate equations in the text to calculate the Arrhenius activation energies at high and low pressures, and the average energy of the reacting molecules at low temperatures. Can you derive an analogous equation for the average energy of the reacting molecules at high pressure?

4.7 Using the rate constants of Fig. 4.5 calculate

(a) The steady-state concentrations of H and CH_3 near the beginning of the reaction.

(b) The steady-state concentrations of H and CH_3 at equilibrium.

(c) The equilibrium constant for the reaction

$$2\,CH_4 \rightarrow C_2H_6 + H_2$$

Compare your results with the concentrations shown in Figure 4.5.

4.8 Using the steady-state assumption, calculate the rate law for ethane decomposition if reactions 1 and 3 are second order (C_2H_6 being the collision partner for reaction 3) and reaction 5 (second-order) is the chief terminating step.

4.9 Activation energies for the six reactions of ethane pyrolysis listed in the text are

Reaction	E (kcal)
1	86
2	11
3	40
4	10
5	0
6	0

Calculate the overall activation energy for ethane pyrolysis in the three typical kinetic situations described in the text, and also that of Problem 4.8.

4.10 It was stated in the text that at about 800°K, in Hinshelwood's temperature range, the steady-state concentrations of O and OH would be much less than that of H. If the ratios are governed by reactions 2, 3, and 4, calculate [O]/[H] and [OH]/[H] for a stoichiometric H_2–O_2 mixture at this temperature, given the following rate constant data, taken from D. M. Baulch, "High Temperature Reaction Rate Data" (The University, Leeds, England, 1968 and 1969):

$$k_2 = 2.2 \times 10^{13}\, e^{-5,150/RT} \quad \text{cm}^3\ \text{mole}^{-1}\ \text{sec}^{-1}$$

$$k_3 = 2.2 \times 10^{14}\, e^{-16,800/RT} \quad \text{cm}^3\ \text{mole}^{-1}\ \text{sec}^{-1}$$

$$k_4 = 1.7 \times 10^{13}\, e^{-9,450/RT} \quad \text{cm}^3\ \text{mole}^{-1}\ \text{sec}^{-1}$$

4.11 At 800°K, the second explosion limit occurs at about 80 Torr in a stoichiometric hydrogen–oxygen mixture. Using the value of k_3 from Problem 4.10 calculate k_5. Compare this with the value

$$k_5 = 6.4 \times 10^{15}\, e^{1000/RT} \quad \text{cm}^6\ \text{mole}^{-2}\ \text{sec}^{-1}$$

from Baulch's tabulation.

5 Chemical Reactions in Solution

Although reactions in solution are more common than gas-phase ones in most laboratories, we have looked at gaseous reactions first because they tend to be simpler. Now, let us consider solutions.

5.1 COMPARISON OF GAS-PHASE AND SOLUTION REACTIONS

In the gas phase, reactant molecules are free of outside influences most of the time, and during reaction interact only with one another. In solution, solvent molecules exert a continuing, usually important, influence on the reactants. Some of the effects that solvent molecules can have are as follows.

5.1.1 Number of Collisions

The number of intermolecular collisions undergone by a reactant molecule in unit time is typically much larger in solution than in the gas phase since the total concentration of molecules is higher, while the speed of translational motion of the molecules is about the same in liquid and gas phases. (The energy needed to vaporize a liquid is used almost entirely to overcome intermolecular attractions, thereby raising the potential energy of the molecules, but not changing their kinetic energy very much.)

In a dilute solution, most of the collisions will be with solvent molecules, and therefore unreactive, but all these collisions will ensure that thermal equilibrium will be maintained, so there will be no problem of fulfilling this requirement of the activated complex theory. The "fall-off" behavior of

Table 5.1 First-order Rate Constants and Activation Energies for the Decomposition of N_2O_5[a]

Solvent	$k \times 10^5$, sec^{-1} at 20°C	E (kcal)
Gas phase	1.65	24.7
Nitrogen tetroxide	3.44	25.0
Ethylidene chloride	3.22	24.9
Chloroform	2.74	24.6
Ethylene Chloride	2.38	24.4
Carbon tetrachloride	2.35	24.2
Pentachloroethane	2.20	25.0
Bromine	2.15	24.0
Nitromethane	~1.50	24.5
Propylene chloride	~0.24	27.0
Nitric acid	~0.05	28.3

[a] From H. Eyring and F. Daniels, *J. Amer. Chem. Soc.* **52,** 1473; 1486 (1930).

unimolecular reactions at low pressures described in Chapter 4 will not be expected at low concentrations in solution, so that as a first approximation we would expect that unimolecular reactions in solution would have rate constants equal to those in the gas phase at high pressures.

Such behavior has been found experimentally. In a classic study reported in 1930, Eyring and Daniels (1) measured the rate of decomposition of N_2O_5 (which is probably a unimolecular reaction) in the gas phase and in several solvents, obtaining the results of Table 5.1. For most of the solvents the solution rate constants are fairly close to that in the gas phase. For the last two, it appears that some special interactions between reactant and solvent must occur to cause the deviations. In a more recent study, Cundall (2) reported that the cis–trans isomerization of dimethyl maleate has about the same Arrhenius A and E values in the gas and liquid phases, while Cain and Solly (3) have made similar observations for the rearrangement

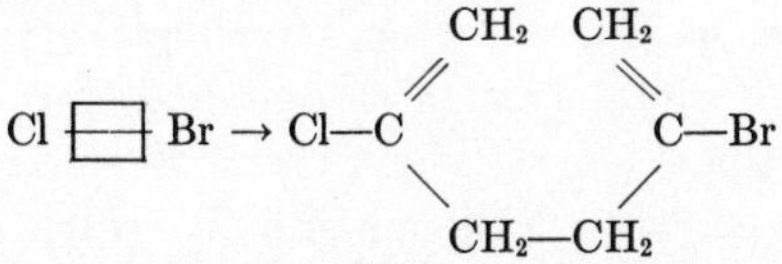

$$\text{Cl}\!-\!\square\!-\!\text{Br} \rightarrow \text{Cl—C}\begin{matrix}\diagup\!\!\diagup \text{CH}_2 \quad \text{CH}_2 \diagdown\!\!\diagdown \\ \diagdown \text{CH}_2\text{—CH}_2 \diagup\end{matrix}\text{C—Br}$$

The presence of the solvent would be expected to increase the number of collisions between two reactant molecules above that calculated by the kinetic theory for the same reactants at the same concentration in an ideal gas state. This comes about because the free volume for molecular motion in a condensed phase is substantially less than the total volume, in contrast to the usual gaseous situation. By the phrase "free volume" we mean the volume occupied by the liquid in excess of the volume it would occupy if it were cooled to 0° K and formed a close-packed crystal lattice. At room temperature the free volume is about 20% of the total molar volume (for a typical nonpolar liquid like CCl_4) and the molecules are somewhat separated, but the distance two adjacent molecules have to travel to collide is considerably less than the center-to-center distance, which is the value used in the kinetic theory calculation. In a very interesting paper published in 1938, Fowler and Slater (4) showed by what they term "sloppy" calculations that the collision rate of reactants in a typical dilute solution will be a few times that in the gas phase.

It has proved difficult to obtain experimental evidence on this point for two reasons. First, there are not many bimolecular reactions that can be studied in both the gas and solution phases. Second, for the few reactions that have been studied, it is not clear that differences between gas-phase and solution rates will be due only to differences in collision number—other interactions between reactants and solvent could be having an effect. Experimental evidence now available suggests that Arrhenius A values in solution are close to those calculated by the simple collision theory, using reasonable values of molecular diameters, in most cases where chemical considerations indicate that solute–solvent interactions will be small.

5.1.2 Pattern of Collisions

The patterns of intermolecular collisions may be substantially different in liquid and gas. The process of diffusion in a liquid normally involves an activation energy since a molecule must squeeze between surrounding molecules to move to another location in the solvent. Its progress in moving around will be much slower than in the gas phase. On the other hand, once it does come in contact with another reactant molecule, it may undergo a number of collisions before diffusing away again (if reaction does not occur), compared to only one collision per approach to another molecule in the gas phase. The net result may be approximately the same number of collisions per second as for the same concentrations in the gas phase (perhaps somewhat more, as suggested above), but the idea of a collision as a definite mechanical event, which is the viewpoint of gas phase kinetics, becomes fuzzy when we consider liquids in which reactant molecules may remain in more or less close contact, within a surrounding, permeable, wall of solvent molecules, for an extended period.

This "solvent cage" effect is the reason one type of reaction that is rare in gases occurs commonly in solution. This is the rearrangement reaction that occurs by elimination and recombination. For example, the process

$$CH_3CH_2CH_2Cl \rightarrow CH_3CHClCH_3$$

is thought to occur in solution by loss of HCl from the reacting molecule, followed by recombination of HCl with the olefin before the HCl has had an opportunity to diffuse away from it. In solution, rearrangement occurs much faster than formation of HCl and propylene, while in the gas phase the reverse is true since there is nothing to hold the intermediate products together once they have formed.

5.1.3 Ions

Ions are encountered much more commonly in solution than in gas-phase reactions. Even substances that are commonly thought of as ionic, such as NaCl, exist primarily as molecules in the gas phase, and it is only by high-energy excitation (such as by electron impact in a mass spectrometer) that ions can be produced in relatively high concentrations in the gas phase.

The reaction $NaCl(s) \rightarrow Na^+(g) + Cl^-(g)$ involves an increase in enthalpy of about 180 kcal, while the comparable reaction in water, $NaCl(s) \rightarrow Na(aq) + Cl^-(aq)$, is almost thermoneutral.

This large difference in energy may be attributed to two factors. First, the dielectric constant of water is high (about 80, compared to 1 for a vacuum) so that the energy necessary to separate two oppositely charged particles is much less in a water environment than in a vacuum. Second, there is a strong tendency for ions to be hydrated in aqueous solution, and this effect further reduces the amount of energy needed to break up a crystal into individual ions.

In Table 5.2 are listed dielectric constants of a few common solvents. It may be seen that the dielectric constant increases with the polar nature of the molecules. Since ionization is so strongly dependent on dielectric constant, kinetic measurements are often made in mixed solvents, such as acetone–water, so that a range of dielectric constants may be used at increments as close as desired.

Hydration (or, in general, solvation) typically occurs most strongly when promoted by electrical forces. Hydration energies for the common inorganic ions are highest for small ions of high charge, such as Al^{3+}, since here the attraction between the negatively charged oxygen atom of water and the positive charge on the ion is strongest. For organic solvents and solutes, solvation will tend to occur when oppositely charged parts of molecules can attract one another, while solvation effects and the formation of ions will be minimal in nonpolar systems.

5.1.4 Solvent Interactions

In liquids it is often necessary to allow for nonideal behavior of reactants due to variations in molecular interactions that cause the ac-

Table 5.2 Dielectric Constants of Selected Solvents at 25°C[a]

Solvent	Dielectric constant
Carbon tetrachloride	2.2
Benzene	2.3
Chloroform	4.8
Ethyl acetate	6.0
Acetic acid	6.2
Chloroacetic acid	12
Ammonia	17
Acetone	21
Ethyl alcohol	24
Methyl alcohol	33
Water	79
Hydrocyanic acid	115

[a] Data from Handbook of Chemistry and Physics," 48th Ed., The Chemical Rubber Company, Cleveland, Ohio, 1967.

tivity to be different from the concentration. This is seldom necessary in gaseous reactions, in which the ideal gas law is a close enough approximation most of the time. Two general types of interaction are found. First, if reactant molecules are solvated in such a way that their reactive parts are substantially affected, then the rate of a reaction may be substantially different than it would be in the gas phase. If a solvent molecule tends to coordinate with a reactant so that the point of reaction with another reactant is covered over, the rate constant for the reaction may be much lower than otherwise expected. Conversely, strong coordination with a solvent molecule, as mentioned above in connection with the formation of ions, may accelerate a dissociation reaction. Second, since electrical forces between ions fall off less rapidly with distance than do other molecular electrical forces, it is found that interionic interactions in solution, among reactant ions, can produce observable effects on reaction rates.

5.2 ACTIVATED COMPLEX THEORY FOR LIQUIDS

In principle, the ACT is as applicable to liquids as to gases. Indeed, the closer approach to energy equilibration caused by the many intermolecular collisions would suggest that ACT would be even better in liquids than in gases. However, the strong intermolecular forces in liquids could be expected to modify the assumed equilibrium between reactants and the activated complex, in some cases to a significant extent, so that quantitative predictions of the theory have not been as dependable in liquid as in gaseous systems. The theory has still been a helpful guide, though, in interpreting experimental results and in predicting how changes in reaction conditions will effect rates. Two such applications will be discussed in this section.

5.2.1 Nonideal Behavior in Solutions

In Chapter 3 we had written the ACT mechanism for a simple bimolecular reaction as

$$\mathrm{A} + \mathrm{BC} \rightarrow \mathrm{ABC} \rightarrow \text{products}$$

where ABC is the activated complex. We had also written an equilibrium constant for the formation of the activated complex

$$K^{\ddagger} = \frac{[\mathrm{ABC}]}{[\mathrm{A}][\mathrm{BC}]} \tag{5.1}$$

For solutions, it would be more appropriate to write

$$K^{\ddagger} = \frac{a_{\mathrm{ABC}}}{a_{\mathrm{A}} a_{\mathrm{BC}}} \tag{5.2}$$

where the a's are activities. If we wish to continue to write rates in terms of concentrations, then we may relate activities to concentrations by the activity coefficient: $a_{\mathrm{A}} = \gamma_{\mathrm{A}}[\mathrm{A}]$, and similarly for the other species. The activity coefficient γ, of course, is the ratio of activity to concentration. Accordingly,

$$K^{\ddagger} = \frac{[\mathrm{ABC}]}{[\mathrm{A}][\mathrm{B}]} \frac{\gamma_{\mathrm{ABC}}}{\gamma_{\mathrm{A}} \gamma_{\mathrm{BC}}} \tag{5.3}$$

If this expression is now used in the derivation of the ACT, as in Section 3.5, the final ACT equation analogous to Eq. (3.70) is

$$k = \frac{\mathbf{k}T}{h} \frac{NQ_{\ddagger}}{Q_{\mathrm{A}} Q_{\mathrm{BC}}} \exp(-E_0/RT) \cdot \frac{\gamma_{\mathrm{A}} \gamma_{\mathrm{BC}}}{\gamma_{\mathrm{ABC}}} \tag{5.4}$$

or

$$k = k^{\circ} \frac{\gamma_{\mathrm{A}} \gamma_{\mathrm{BC}}}{\gamma_{\mathrm{ABC}}} \tag{5.5}$$

where k° is the rate constant in the standard state for which the γ's are 1.

This equation can be used to relate rate constants in solution to those in the gas phase, but even more commonly it is used to relate rate constants under different liquid-phase conditions. For example, the γ's of electrolytes in dilute solution may be predicted by the Debye–Hückel theory, and this relationship may be used to correlate data at different concentrations, $k°$ being taken for infinitely dilute solution (by extrapolation) at which the γ's approach 1. Let us develop the appropriate relationship.

According to the Debye–Hückel theory, activities of electrolytes in dilute aqueous solution at room temperature are given by the equation

$$\log \gamma = -0.51 z^2 \mu^{1/2} \qquad (5.6)$$

where the value 0.51 is mainly a collection of physical constants and is proportional to $(\epsilon T)^{-3/2}$, where ϵ is the dielectric constant of the solution, z is the charge on the ion in electron units, and μ is the ionic strength, a measure of the effects of all the ions in solution. Specifically,

$$\mu = \tfrac{1}{2} \sum c_i z_i^2 \qquad (5.7)$$

where c_i is the concentration of a given ion, and z is again the ionic charge. For example, in a 0.1 M solution of $BaCl_2$, if Ba^{2+} is called ion 1 and Cl^- ion 2,

$$\mu = \tfrac{1}{2}[0.1 \times 2^2 + 0.2 \times (-1)^2] = 0.3$$

According to Eq. (5.6), γ will be less than 1 for all finite ionic concentrations.

If Eq. (5.5) is rewritten as

$$\log k = \log k° + \log \gamma_{\mathrm{A}} + \log \gamma_{\mathrm{BC}} - \log \gamma_{\mathrm{ABC}} \qquad (5.8)$$

we can introduce Eq. (5.6) to obtain

$$\log k = \log k° - 0.51 \mu^{1/2} [z_{\mathrm{A}}^2 + z_{\mathrm{BC}}^2 - z_{\mathrm{ABC}}^2] \qquad (5.9)$$

and then recognize that these z's are not independent, but that

$$z_{\mathrm{ABC}} = z_{\mathrm{A}} + z_{\mathrm{BC}}$$

so that

$$z_{\mathrm{ABC}}^2 = (z_{\mathrm{A}} + z_{\mathrm{BC}})^2 = z_{\mathrm{A}}^2 + 2 z_{\mathrm{A}} z_{\mathrm{BC}} + z_{\mathrm{BC}}^2$$

Therefore,

$$\log k = \log k° - 0.51 \mu^{1/2} [z_{\mathrm{A}}^2 + z_{\mathrm{BC}}^2 - z_{\mathrm{A}}^2 - 2 z_{\mathrm{A}} z_{\mathrm{BC}} - z_{\mathrm{BC}}^2]$$

$$\log k = \log k° + 1.02 z_{\mathrm{A}} z_{\mathrm{BC}} \mu^{1/2} \qquad (5.10)$$

This equation, which has been tested for several ionic systems, indicates that rate constants for similarly charged particles will increase with concentration, while rates for oppositely charged ions will decrease with concentration. A look back at Eq. (5.3) will show that this is due to the effect of the γ's on [ABC] at the steady-state conditions.

Equation (5.5) can be used whenever we have a way of calculating (or estimating) activity coefficients of reactants and the activated complex, and when we have a rate constant at some known condition of activities.

5.2.2 Thermodynamic Formulation of the Activated Complex Theory

A formulation commonly used in discussions of kinetics in solution is obtained from Eq. (3.70) by noting that this equation can be written

$$k = \frac{\mathbf{k}T}{h} K_{\ddagger} \tag{5.11}$$

where $K_{\ddagger}$ is a special equilibrium constant for the reaction of formation of activated complexes from reactants—it is "special" because the contribution of the reaction coordinate of the activated complex is not included in the partition function $Q_{\ddagger}$ for the complex. In the bimolecular formulation of Eq. (3.70),

$$K_{\ddagger} = \frac{NQ_{\ddagger}}{Q_{\mathrm{A}}Q_{\mathrm{BC}}} \exp\left(-\frac{E_0}{RT}\right) \tag{5.12}$$

while for a unimolecular reaction, $K_{\ddagger}$ would be

$$K_{\ddagger} = \frac{Q_{\ddagger}}{Q_{\mathrm{A}}} \exp\left(-\frac{E_0}{RT}\right) \tag{5.13}$$

From thermodynamics, we may write

$$\Delta G^\circ = -RT \ln K \tag{5.14}$$

where ΔG° is the free energy change at standard conditions when 1 mole of reactant changes to one mole of product, or

$$K = \exp\left(-\frac{\Delta G^\circ}{RT}\right) \tag{5.15}$$

Moreover, thermodynamics gives

$$\Delta G = \Delta H - T\,\Delta S \tag{5.16}$$

where ΔH is the enthalpy (heat content) change in the reaction, and ΔS is the entropy change. Therefore,

$$K = \exp\left(-\frac{\Delta H^\circ - T\,\Delta S^\circ}{RT}\right) = \exp\left(\frac{\Delta S^\circ}{R}\right) \exp\left(-\frac{\Delta H^\circ}{RT}\right)$$

and we can write a "thermodynamic" form of the rate constant as

$$k = \frac{\mathbf{k}T}{h} \exp\left(\frac{\Delta S_{\ddagger}^{\circ}}{R}\right) \exp\left(-\frac{\Delta H_{\ddagger}^{\circ}}{RT}\right) \tag{5.17}$$

where $\Delta G_{\ddagger}^{\circ}$, $\Delta S_{\ddagger}^{\circ}$, and $\Delta H_{\ddagger}^{\circ}$ are commonly called the free energy, entropy, and enthalpy of activation. We must keep in mind the ways in which these quantities have been derived, not letting ourselves suppose that they refer to ordinary thermodynamic quantities that define a real equilibrium constant.

$\Delta S_{\ddagger}^{\circ}$ and $\Delta H_{\ddagger}^{\circ}$ are temperature dependent but, as for ΔS° and ΔH° of ordinary reactions, they usually change slowly with temperature, and in a typical kinetic study extending over a range of 10 to 100 in rate constant, data can be interpreted in terms of average values of $\Delta S_{\ddagger}^{\circ}$ and $\Delta H_{\ddagger}^{\circ}$.

If the experimental results are first expressed in the customary Arrhenius form, $\Delta S_{\ddagger}^{\circ}$ and $\Delta H_{\ddagger}^{\circ}$ can be obtained by setting

$$Ae^{-E/RT} = \frac{\mathbf{k}T}{h} \exp\left(\frac{\Delta S_{\ddagger}^{\circ}}{R}\right) \exp\left(-\frac{\Delta H_{\ddagger}^{\circ}}{RT}\right) \tag{5.18}$$

and using the approach of Section 2.8 to obtain

$$\Delta H_{\ddagger}^{\circ} = E - RT$$

$$\Delta S_{\ddagger}^{\circ} = R\left(\ln A - \ln \frac{\mathbf{k}T}{h} - 1\right) \tag{5.19}$$

where T is the average of the experimental range.

What sorts of numbers do we obtain for these "thermodynamic" quantities, particularly for $\Delta S_{\ddagger}$, since $\Delta H_{\ddagger}$ is simply related to the Arrhenius E? For the "normal" first order, probably unimolecular decomposition of N_2O_5 in carbon tetrachloride, the date of Table 5.1 give an Arrhenius equation

$$k = 2.6 \times 10^{13} e^{-24,200/RT}$$

near room temperature. If we take $T = 300°$ K (293° would be just as good a choice), then

$$\Delta S_{\ddagger}^{\circ} = 1.987 \left[2.303 \log(2.6 \times 10^{13}) - 2.303 \log \frac{1.38 \times 10^{-16} \times 300}{6.6 \times 10^{-27}} - 1\right]$$

$$= 1.987[30.90 - 29.48 - 1.00] = 0.8 \quad \text{cal deg}^{-1}\ \text{mole}^{-1}$$

Values of $\Delta S_{\ddagger}°$ close to zero are typical of unimolecular reactions where interactions with the solvent are small or, more generally, where the interaction of the solvent with both reactant molecule and activated complex will be expected to be about the same.

Let us look also at a bimolecular reaction. In the 1930s Fairclough and Hinshelwood (5) found that for the reaction

$$C_2H_5Br + OH^- \rightarrow C_2H_5OH + Br^-$$

in a water–acetone solution at an average temperature of 325° K,

$$k = 2.9 \times 10^9 \, e^{-18,700/RT} \quad \text{mole}^{-1}\,\ell\,\text{sec}^{-1}$$

From Eq. (5.20) we obtain $\Delta S_{\ddagger}° = -14.8$ cal deg^{-1} mole^{-1}, a rather substantial negative value. To some extent the negative value can be understood because the combination of two free species to form a single activated complex molecule represents an increase in ordering, and hence a decrease in entropy. However, two points should be considered. First, the entropy contribution of the reaction coordinate has not been included in $\Delta S_{\ddagger}°$, and this could be fairly large since the reaction coordinate has been thought to be a low-frequency vibration. Accordingly, $\Delta S°$ for the process of forming an activated complex from reactants will not be as negative as $\Delta S_{\ddagger}°$. Secondly, the numerical value of $\Delta S_{\ddagger}°$ depends on the standard state, which is of course 1 mole/ℓ in this example so far. If units of mole^{-1} cc sec^{-1} are used, then

$$k = 2.9 \times 10^{12} e^{-18,700/RT} \quad \text{mole}^{-1}\,\text{cc sec}^{-1}$$

$\Delta S_{\ddagger}°$ becomes -1.1 cal deg^{-1} mole^{-1} for a standard state of 1 mole^{-1} cc sec^{-1}.

The reason $\Delta S_{\ddagger}°$ depends on the standard concentration can be appreciated if we look at the entropy change when a solution is diluted. When two fluids mix, the entropy change due to the increase in disorder is, for 1 mole of total substance,

$$\Delta S = -R(x_1 \ln x_1 + x_2 \ln x_2) \tag{5.20}$$

where x_1 and x_2 are the mole fractions of the two substances after mixing. For 1 mole of substance 2 (the reactant, with substance 1 being the solvent), the increase in entropy per mole is $-R \ln x_2$. Accordingly, the more dilute a solution is, the greater the entropy per mole of solute, the logarithmic dependence on concentration being analogous to the logarithmic dependence on pressure for gases. Accordingly when we dilute a solution from 1 mole/cc to 1 mole/ℓ, the entropy of *each* reactant increases by $R \ln 10^3$, while the entropy of the activated complex also increases by the same amount, so the net change in $\Delta S_{\ddagger}°$ is $R \ln 10^3 - 2(R \ln 10^3) = -R \ln 10^3$ or -13.7 cal deg^{-1} mole^{-1}.

5.3 FURTHER APPLICATION OF ACT TO IONIC REACTIONS

In the previous section we looked at the Debye–Hückel formulation for the effect of the total ionic environment on ionic reactions. Here we will develop an expression for the effect of the charges on the reacting ions on the rate constant, in a very dilute solution where the two reacting ions are far from any others. We will attack the problem by calculating the free energy change when the ions come together to form the activated complex.

Consider two ions of charges z_1e and z_2e esu each, where z_1 and z_2 are the numbers of electron charges on each ion (z is negative if the ion is negatively charged) and e is the absolute value of the electronic charge, in coulombs. The electrical force between the ions is $z_1z_2e^2/\epsilon r^2$ dyne, where ϵ is the dielectric constant of the solvent and r the interionic distance. To bring the ions from far apart to a certain small distance apart r_c, at which small distance the complex would form, would require work $= \int$ force $\times$ distance. In order to show positive work done on the ions when they have charges of the same sign, we will write

$$\text{work} = -\int_{\infty}^{r_c} \frac{z_1z_2e^2}{\epsilon r^2}\,dr = \frac{z_1z_2e^2}{\epsilon r_c} \tag{5.21}$$

Now, electrical work done on a system corresponds to a free energy increase, so we can say that this electrical work introduces an additional multiplicative factor into the thermodynamic rate constant expression:

$$k = \frac{\mathbf{k}T}{h}\exp\left(-\frac{\Delta G_{\ddagger}{}^{\circ}}{RT}\right)$$

$$= \frac{\mathbf{k}T}{h}\exp\left(-\frac{\Delta G_{\ddagger}{}^{\circ}}{RT}\right)_{\text{no charge}}\exp\left(-\frac{\Delta G_{\ddagger}{}^{\circ}}{RT}\right)_{\text{electrical}} \tag{5.22}$$

Let us evaluate these equations for water ($e = 79$ at 298° K), assuming z_1 and z_2 are each $+1$ and r_c is 3 Å (a reasonable value, the actual values depending on the configuration of the activated complex for a particular reaction). Per ion pair, then

$$\text{electrical work} = \frac{(4.8 \times 10^{-10})^2}{79 \times 3 \times 10^{-8}} = 9.7 \times 10^{-14} \quad \text{erg}$$

or, on a mole basis

$$\Delta G_{\ddagger}{}^{\circ}{}_{\text{electrical}} = \frac{9.7 \times 10^{-14} \times 6 \times 10^{23}}{4.18 \times 10^{7}} = 1400 \quad \text{cal}$$

so that from (5.22),

$$k = k_{\text{no charge}}\, e^{-1400/1.987\times 298} = 0.095 k_{\text{no charge}}$$

On the other hand, if the charges had been of opposite signs, the effect would have been to increase k by $e^{1400/RT}$, or by a factor of 10.5. For more highly charged ions, the effects of the charges will be correspondingly greater, and the effects will, of course, be larger in solutions of lower dielectric constant.

We can go further in our analysis by breaking down the electrical effect into its individual effects on the entropy and enthalpy of activation. From thermodynamics,

$$\Delta S = -(\partial \Delta G/\partial T)_P \tag{5.23}$$

In the expression for the free energy change (Eq. (5.21)) the only temperature-dependent part is ϵ, the dielectric constant. Thus,

$$\Delta S_{\ddagger}{}^{\circ}{}_{\text{electrical}} = \frac{z_1 z_2 e^2}{\epsilon^2 r_c} \frac{d\epsilon}{dT} \quad \text{erg molecule}^{-1}\ \text{deg}^{-1} \tag{5.24}$$

Near 25° C, $d\epsilon/dT$ is approximately $-0.3\ \text{deg}^{-1}$, so that, if z_1 and z_2 are 1, $\epsilon = 79$ and $r_c = 3 \times 10^{-8}$ cm as in our previous example, then

$$\Delta S^{\circ}{}_{\text{electrical}} = \frac{-(4.8 \times 10^{-10})^2}{79^2 \times 3 \times 10^{-8}} \times 0.3 = -3.7 \times 10^{-12} \quad \text{erg deg}^{-1}$$

or, on a mole basis,

$$\Delta S_{\ddagger}{}^{\circ}{}_{\text{electrical}} = -5.3\ \text{cal deg}^{-1}$$

Since

$$\Delta G = \Delta H - T\,\Delta S$$

we may write

$$\Delta H_{\ddagger}{}^{\circ}{}_{\text{electrical}} = \Delta G_{\ddagger}{}^{\circ}{}_{\text{electrical}} + T\,\Delta S_{\ddagger}{}^{\circ}{}_{\text{electrical}}$$

$$= 1400 - 298 \times 5.1 = -180 \quad \text{cal/mole}$$

in our present example. It appears that the charges affect the rate constants primarily through the entropy term. In a medium of lower dielectrical constant, the relative effects on $\Delta S_{\ddagger}{}^{\circ}$ and $\Delta H_{\ddagger}{}^{\circ}$ would change.

5.4 LINEAR FREE ENERGY RELATIONSHIPS

In 1935 Hammett reported (6) a generalization that has systemized a great deal of kinetic data on organic reactions in solution, and that has been a very good predictive method for estimating rate data where none are available. In working out his generalization, he received a valuable hint from the ACT idea that the rate constant for a reaction depends on the equilibrium constant for the formation of the activated complex. This idea led him to compare the effects of changing substituents in the benzene ring of a series of benzoic acids on (a) the ionization constants of the acids and (b) the rates of hydrolysis of esters of the acids. On plotting the logarithms of the rate constants versus the logarithms of the ionization constants, he obtained straight lines.

In Fig. 5.1 a number of sets of data are plotted in this way, the figure being taken from a paper by Burkhardt *et al.* (7) published in 1936. Several features of these curves are worth noting. First, the choice of using log K for the ionization of the benzoic acids as abscissa is an arbitrary one, which Hammett recommended because accurate, extensive data were available. Any one of the reactions could have been used as a basis against which to plot the others, while still maintaining the linear relationships. However, Hammett's original choice was a good one and is still used. Second, the slopes of the straight lines are quite variable, indicating that the substituents affect the reactions in different ways. For curves with a slope near 1 (such as curves B and E) the substituents have about the same effects on the rates as they do on ionization. Curves with negative slopes correspond to opposite effects of substituents on the two reactions being compared. Third, we should note that the absolute values of the rate constants are not correlated, but only the changes produced by the substituents.

In his monograph (8) published in 1940, Hammett presented an extensive compilation of rate and equilibrium data on substituted aromatic compounds, along with a very convenient algebraic way of presenting all the data. Each series of rate (or equilibrium) data could be represented by an equation

$$\log k \text{ (or } K) = \log k_0 \text{ (or } K_0) + \rho\sigma \tag{5.25}$$

where k_0 (or K_0) represents the rate constant (or equilibrium constant) for the reaction involving the unsubstituted compound, ρ is a constant characteristic of each reaction, and σ is a constant characteristic of each substituent. For the standard reaction (ionization of benzoic acids) ρ is taken to be 1.

Let us see how these various quantities could actually be obtained. We would start by measuring the dissociation constant of benzoic acid (choosing a particular solvent, water, at 298° K) which would give us K_0 for that reaction. Then, we would measure the dissociation constants of substituted benzoic acids (such as *p*-nitrobenzoic acid) and thereby obtain the substituent constants for each substituent used:

$$\sigma = \log K - \log K_0 \tag{5.26}$$

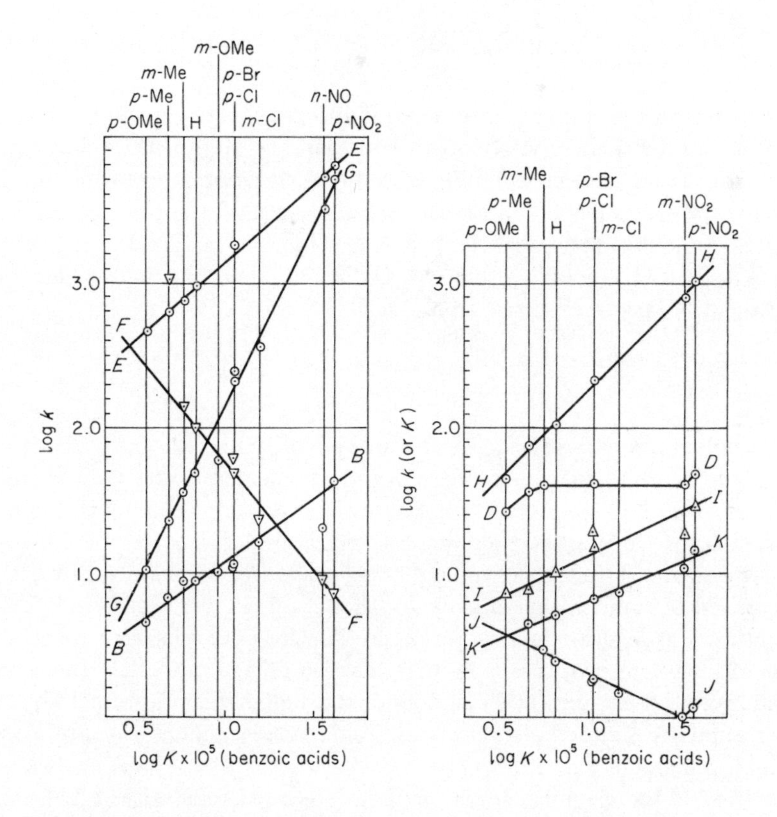

Ordinates for curves:

B log($k \times 10^5$) for acid-catalyzed hydrolysis of the substituted potassium phenyl-sulfates at 48.6°
E log($k \times 10^4$) for alkaline hydrolysis of the substituted benzamides at 100°
G log($k \times 10^4$) for alkaline hydrolysis of the substituted ethyl benzoates at 30°
F log k for hydrolysis of the substituted benzyl chlorides by aqueous alcohol at 83°
H log($k \times 10^3$) for alkaline hydrolysis of the substituted ethyl cinnamates at 30°
D log($k \times 10^4$) for acid hydrolysis of the substituted benzamides at 100°
I log($k \times 10^3$) for reaction KI + C_6H_4X S CH_2 $_2$ Cl in acetone at 55°
J log($k \times 10$) for bromination of substituted acetophenones at 25°
K log($K \times 10^5$) of the substituted phenyl-acetic acids at 25°

Abscissae: logarithms of the dissociation constants ($\times 10^5$) of the substituted benzoic acids.

Fig. 5.1 Several plots of rate and equilibrium constants for substituted aromatic compounds versus the ionization constants for the substituted benzoic acids (7). See original source for attributions for data, omitted here.

since for this reaction ρ is 1. One additional measurement is needed, in principle, for each type of substituent. The position of the substituent in the ring is important, different constants being obtained for ortho, meta, and para positions. To obtain the necessary data for a given reaction, two measurements are enough in principle. One would find the rate for the unsubstituted reactant, which would be k_0, then the rate when a substituent of known σ is used, from which

$$\rho = \frac{\log k - \log k_0}{\sigma} \tag{5.27}$$

Actually, for many reactions data are available for a variety of substituents, and the best way to make tables of ρ and σ values is to carry out a regression analysis of all possible data. On making such a study, Hammett found that substituent constants could be set nearly independently of solvent or temperature, the effects of solvent being accounted for mainly by variations in ρ and of temperature mainly by the k_0 values. He also found that his method was less satisfactory for ortho than for para and meta substituents in the benzene ring, perhaps due to steric effects, and that the ρ and σ values found for aromatic compounds did not predict rates and equilibria for aliphatic compounds very well. Some typical values of ρ are given in Table 5.3 and of σ in Table 5.4.

What is the chemical basis for these observed regularities? Let us look first at the ionization constants of the acids. The dissociation constant of

Table 5.3 Hammett Reaction Constants ρ

Reaction	E (equilibrium) or R (rate)	ρ
Dissociation of benzoic acids in water at 25°C	E	1.00
Dissociation of benzoic acids in ethanol at 25°C	E	1.96
Dissociation of phenols in water at 25°C	E	2.11
Base hydrolysis of ethyl benzoates in 85% ethanol solution at 50°C	R	2.12
Acid hydrolysis of ethyl benzoates in 60% ethanol solution at 100°C	R	0.14
Base hydrolysis of benzyl chlorides in water at 30°C	R	−0.33
Reaction of $Ar^1Ar^2CHCl + CH_3OH$ to form methyl ester, in CH_3OH at 25°C	R	−4.67

Values from H. H. Jaffe (*Chem. Rev.* **53,** 191 (1953)), who lists 218 reactions altogether.

Table 5.4 Substituent Constants σ

Group	σ (meta)[a]	σ (para)[a]	σ_I[b]
O^-	−0.71	−0.52	−0.12
CH_3	−0.07	−0.17	−0.05
H	0.00	0.00	0.00
OCH_3	0.12	−0.27	0.25
OH	0.12	−0.37	0.25
Cl	0.37	0.06	0.47
Br	0.39	0.23	0.45
CN	0.56	0.66	0.56
NO_2	0.71	0.78	0.63

[a] Meta and para values from D. H. McDaniel and H. C. Brown, *J. Org. Chem.* **23,** 420 (1958) and G. B. Barlin and D. D. Perrin, *Quart. Rev. Chem. Soc.* **20,** 75 (1966).
[b] σ_I values from R. W. Taft and I. C. Lewis, *J. Amer. Chem. Soc.* **80,** 2436 (1957); R. W. Taft, E. Price, I. R. Fox, I. C. Lewis, K. K. Anderson, and G. T. Davis, *J. Amer. Chem. Soc.* **85,** 709 (1963).

an acid will depend on the electron distribution between the acid hydrogen and the oxygen atom to which it is attached. The moderate dissociation constant of benzoic acid in water (6×10^{-5}) indicates that in the undissociated acid the pair of electrons bonding the H and O atoms is somewhat displaced toward the oxygen. We would expect that any substituent in the benzene ring that would tend to attract electrons would cause further displacement of the electrons holding the hydrogen, and cause the dissociation constant of the substituted acid to be greater than that of the unsubstituted one. In terms of the Hammett equation, electron-attracting substituents in the benzene ring would have positive σ values, and vice versa. Examination of Table 5.4 shows that, indeed, groups such as Cl and NO_2 which we would think of as strongly electronegative, do have high σ values, groups like CH_3 that have little tendency either to withdraw or to donate electrons have σ values close to zero, while the one example of a negatively charged substituent (O^-) has a large negative σ.

In general it is assumed that the effect of substituents on other equilibria and reaction rate constants are due to changes in electron density in the reaction center. Any other process that is increased to the same extent as the ionization of benzoic acids by the withdrawal of electrons from the reaction center wall have a ρ of 1. If the reaction is more sensitive in the same sense, ρ will be greater than 1. If the reaction is hindered by the withdrawal of electrons, its ρ value will be less than 1, and perhaps negative.

In the second edition of his book (8), Hammett has pointed out that even in the two cases of meta and para substituents in the benzene ring, there are many complications that may cause data points to lie off correlation curves such as Fig. 5.1, or conversely, may cause predictions based on the equation to be somewhat incorrect. One of the chief complications is resonance, which tends to be especially common for para substituents. Taft and co-workers (9–11) tried to develop a series of σ values that depended only on the electron withdrawing ability (inductive effect) of the groups, and not on resonance. These values, presumably, would be more characteristic of the groups than the values Hammett had found, and in particular would be usable in predicting reaction rates and equilibria in aliphatic systems, where resonance effects are minimal.

In his first papers, Taft developed an equation analogous to Hammett's:

$$\log \frac{k}{k_0} \left(\text{or} \frac{K}{K_\circ} \right) = \rho^* \sigma^* \tag{5.28}$$

where ρ^* and σ^* are reaction and substituent constants as before, but particularly adapted to aliphatic systems. Knowing that steric effects are more important in aliphatic than in aromatic systems, he decided to choose as his standard reaction the hydrolysis of ethyl esters, a reaction that is catalyzed both by acids and by bases, and can be studied in both acid and basic solutions. He considered that the steric effects would be nearly the same for both types of catalysis since the H atoms in H_3O^+ and OH^- are almost embedded in the electron cloud of the oxygen, and tend to produce minimal steric effects, and that a polar (or inductive) substituent constant σ^* could be defined by the equation

$$(\rho_B^* - \rho_A^*)\sigma^* = \left(\log \frac{k}{k_0} \right)_B - \left(\log \frac{k}{k_0} \right)_A \tag{5.29}$$

where $\rho_B^* - \rho_A^*$ is the difference in the reaction constants for basic and acid hydrolysis of the ester, and the B and A subscripts refer to ratios of rate constants with and without substituents for basic and acid catalysis respectively. The assumption that $(\rho_B^* - \rho_A^*) = 1$ enabled Taft to calculate σ^* values (using CH_3 as the standard substituent with $\rho^* = 0$, rather than H) and ρ^* values for other reactions, which he tabulated in the last paper of Reference 9. We have not listed these values here because, due to the importance of steric effects, these constants do not have the same predictive usefulness as Hammett's constants.

Later (10, 11) Taft reformulated his ρ^* values to make them more comparable (for correlative purposes) with Hammett's data. First, he introduced a CH_2 group into his reacting molecules, thus redefining the substituents. For example, in an esterification reaction, he went from

$$R - COOH + HOC_2H_5 \rightarrow \text{ester}$$

where R is the substituent, to

$$X - CH_2 - COOH + HOC_2H_5 \rightarrow \text{ester}$$

This change placed the substituent in a relationship to the reaction center which is more analogous to the Hammett situation, and also converted the "standard" substituent back to H. Second, since (as indicated in Table 5.3) the difference in $\rho_B - \rho_A$ is probably closer to 2 than 1, Taft examined available data and concluded that the polar substituent values could be made comparable to the Hammett σ values if they were multiplied by 1/2.22 or 0.45. The new values are symbolized by ρ_I, or inductive substituent constants, and some are shown in Table 5.4. A comparison of ρ_I with the meta and para σ gives an indication of the contribution of factors other than the inductive effect to the overall effect of a substituent.

Efforts to explain variations in reaction rates in terms of inductive, steric, and resonance effects, and to develop more complicated equations of the Hammett type that account for these effects, have stimulated a great number of experimental kinetic studies, and have produced a much greater understanding of the mechanisms of organic reactions. As Hammett points out in the second edition of his book (8) this work is continuing.

5.5 DIFFUSION—CONTROLLED REACTIONS

Many of the common rapid reactions of inorganic solution chemistry appear to be limited in speed only by the rate at which the species can get together to react. In an acid–base titration, for example, the indicator color changes appear to the analyst to occur as soon as the solutions mix. The gross convective mixing of a few drops of reagent added to a flask of solution frequently takes a few seconds, but beyond that the reaction seems to go very fast indeed.

If it is assumed that reagent molecules could be thoroughly mixed without reaction, and that at some point in time reaction could start as soon as reactant molecules diffuse into contact with one another, with no (or negligible) activation energy, then the rate of reaction can be calculated if some fairly reasonable assumptions are made. Essentially, these are that Fick's law for diffusion is obeyed—the rate of diffusion is proportional to the concentration gradient—and that the Stokes–Einstein law of diffusion, which was derived for relatively large spheres moving in a homogeneous medium as in Millikan's oil drop experiment, can be applied to molecules and ions to relate the rate of diffusion to the viscosity of the liquid. With these assumptions, the equation

$$k_\mathrm{D} = \left(\frac{8RT}{3000\ \eta}\right)\left(\frac{\delta}{e^\delta - 1}\right) \quad \text{mole}^{-1}\ \ell\ \text{sec}^{-1} \tag{5.30}$$

is obtained, where R is in units of erg deg^{-1} mole^{-1} and η is the viscosity of the liquid at temperature T, in units of poises (the poise is a cgs with dimensions g cm^{-1} sec^{-1}). The term δ accounts for the presence of charges on the particles, and is given by

$$\delta = \frac{z_\mathrm{A} z_\mathrm{B} e^2}{\epsilon \mathbf{k} T(r_\mathrm{A} + r_\mathrm{B})} \tag{5.31}$$

where z_A and z_B are the charges, in electron units, on the ions, e is the electronic charge, ϵ the dielectric constant, and r_A and r_B the ionic radii. As you may have guessed, this part of the relationship was derived by Debye (12).

The first part of the rate constant equation may be easily evaluated for water, with $\eta = 0.01$ P, at 300° K, as

$$k_\mathrm{D}\ (\text{for uncharged particles}) = \frac{8 \times 4.18 \times 10^7 \times 300}{3000 \times 0.01}$$

$$= 0.32 \times 10^{10} \quad \text{mole}^{-1}\ \ell\ \text{sec}^{-1}$$

To evaluate δ, we must make assumptions as to r_A and r_B, as well as know the charges on the ions and the dielectric constant of the medium. If we take $r_\mathrm{A} + r_\mathrm{B} = 3 \times 10^{-8}$ cm and $\epsilon = 79$ (as for water at room temperature), then

$$\delta = 2.4 z_\mathrm{A} z_\mathrm{B}$$

where, of course, the value 2.4 is not very accurately known. For charges of +1 and −1, as for a neutralization of a strong acid with a strong base, the electrical contribution would be $-2.4/(e^{-2.4} - 1) \sim 2.6$, so a moderate increase in rate would be expected due to the electrical forces.

Bimolecular rate constants of 10^{10} mole^{-1} ℓ sec^{-1} are very high compared to those we have considered so far, at least in the liquid phase, in which kinetic measurements have mostly been carried out on a time scale of hours by periodically sampling a reaction mixture. It has been only recently that special methods have been developed to measure these very fast processes. One idea that has been used in several types of apparatus (discussed further in Chapter 7) is the so-called "relaxation method."

In the application of relaxation methods the physical impracticality of adding two highly reactive solutions to one another and mixing them thoroughly in a few microseconds is recognized. Accordingly, the solutions are mixed in a leisurely fashion and allowed to come to equilibrium, and then a sudden change in pressure, temperature, or some other quantity is brought about in the solution; this causes the system to suddenly not be at equilibrium. The process of return to equilibrium via the diffusion-controlled reaction is then followed using fast-responding instruments.

As an example, suppose we consider the dissociation of an acid, symbolized generally as an elementary process

$$A \underset{k_-}{\overset{k}{\rightleftarrows}} B + C$$

where the concentrations of A, B, and C at equilibrium are a, b, and c. Therefore we can write

$$K = \frac{k}{k_-} = \frac{bc}{a}$$

If we assume that at the beginning of the experiment (immediately after a sudden rise of temperature, for example, which means that the former equilibrium concentrations are no longer equilibrium concentrations at the new temperature) there is a small excess of [A] over the equilibrium value, let us symbolize this excess by x, so that the actual concentration of A will be $a + x$. By stoichiometry, the actual concentrations of B and C will be $b - x$ and $c - x$.

We will expect that x will diminish during the experiment, approaching zero eventually. At some time

$$\begin{aligned} \frac{dx}{dt} &= -k(a + x) + k_-(b - x)(c - x) \\ &= ka - kx + k_-bc - k_-bx - k_-cx + k_-x^2 \qquad (5.32) \\ &= -ka + k_-bc - x(k + k_-b + k_-c) + k_-x^2 \end{aligned}$$

This equation can be simplified considerably. The first two terms add up to zero, since ka and k_-bc are the forward and reverse rates at equilibrium. The last term will probably be very small compared to the second since in most of these experiments x will be small compared to a, b, and c, and we will neglect it. Finally, since the equilibrium constant is probably known (or can be measured since the system comes rapidly to equilibrium), we can replace k_- by k/K. Accordingly,

$$\frac{dx}{dt} = -kx\left(1 + \frac{b}{K} + \frac{c}{K}\right)$$

which can be integrated between the limits of x_0 and x to give

$$x = x_0 \exp\left(-k\left(1 + \frac{b}{K} + \frac{c}{K}\right)t\right) \tag{5.33}$$

This exponential curve is just like that found for a first-order reaction with a rate constant equal to $k(1 + b/K + c/K)$; similar equations would be found for other rate laws. The half-life would be

$$t_{1/2} = \frac{0.693}{k(1 + b/K + c/K)}$$

so that

$$k = \frac{0.693}{t_{1/2}(1 + b/K + c/K)} \tag{5.34}$$

This equation emphasizes the fact that the method of analysis need not be calibrated to give absolute values of concentrations. All that is necessary is to measure on a relative scale the concentration of one of the reactants or products. The analytical method should be fairly sensitive, however, since x will normally be small. It is also not usually necessary to measure precisely the extent of the change of conditions of the reaction system since the equilibrium constant and the concentrations can be measured at the end of the experiment, which is the condition for which they were defined in the derivation.

It has been customary in reporting data obtained by relaxation methods to give the "relaxation time," denoted by the Greek letter τ and signifying the time required for x to drop to $1/e$ of its initial value, rather than $t_{1/2}$.

Table 5.5 Some Fast Reactions in Solution[a]

Reaction	k (mole^{-1} l sec^{-1}) at room temperature
$H_3O^+ + OH^- \rightarrow 2\ H_2O$	1.3×10^{11}
$H_3O^+ + C_6H_5COO^- \rightarrow C_6H_5COOH$	4×10^{10}
$NH_4^+ + OH^- \rightarrow NH_4OH$	3×10^{10}
$H_3O^+ + H_2O \rightarrow H_2O + H_3O^+$	1×10^{10}
$CH_3OH + OH^- \rightarrow CH_3O^- + H_2O$	3×10^{6}
$Cd(CN)_3^- + CN^- \rightarrow Cd(CN)_4^{-2}$	7×10^{7}

[a] Data from F. E. Caldin, "Fast Reactions in Solution." Blackwell, Oxford, 1964.

τ is, therefore, longer than $t_{1/2}$ by the factor 1/0.693. From Eq. (5.33) it can be seen that

$$k\left(1 + \frac{b}{K} + \frac{c}{K}\right)\tau = 1$$

or

$$k = \frac{1}{\tau(1 + b/K + c/K)} \tag{5.35}$$

In general, if the apparent first-order rate constant for the change in x with time is k_a, then $\tau = 1/k_a$.

Rate constants obtained by this method for some familiar diffusion-controlled reactions are listed in Table 5.5.

5.6 REACTIONS CATALYZED BY ENZYMES

Enzymes are large protein molecules, with molecular weights typically in the range 10^4–10^6, that are highly effective catalysts for a wide variety of reactions of biological importance. The size of enzymes, particularly by comparison with the relatively small molecules whose reactions are catalyzed, leads to the question as to whether these processes should be called solution reactions or heterogeneous ones. Any distinction is arbitrary, and indeed enzyme reactions have some characteristics of both types. Since enzymes have dimensions of the order of 100 Å, which is too small to scatter much visible light, it is not unreasonable to treat them as big molecules in solution.

Enzymes are generally assumed to have one or more small regions, or active sites, at which reactants are absorbed, and catalytic activity occurs. However, the rest of the molecule apparently makes some contribution to the overall effect since attempts to produce enzymelike activity by smaller molecules with configurations similar to the active site configurations of enzymes have not been successful.

The basic kinetic equations for enzyme activity were developed over a period of years by Henri (13), Michaelis and Menten (14), and Briggs and Haldane (15), over the period 1902–1925. The particular equations for any given enzyme will depend, of course, on the type of reaction being catalyzed. We will follow the usual approach of writing down the way in which an enzyme would be expected to catalyze a simple unimolecular reaction

$$\mathrm{S} \rightarrow \mathrm{P}$$

where S is the reactant molecule, usually called the *substrate* molecule by enzyme chemists, and P is the product of the reaction. In the simplest formulation, we can think of the reaction occurring in two stages:

$$\mathrm{E} + \mathrm{S} \underset{k_{-1}}{\overset{k_1}{\rightleftarrows}} \mathrm{ES}$$

$$\mathrm{ES} \overset{k_2}{\rightarrow} \mathrm{E} + \mathrm{P} \tag{5.36}$$

where ES corresponds to the combination of an enzyme and an (unreacted) substrate molecule. Of course, as catalysts, enzymes promote reaction in both directions, but in most biological systems supply of reactant and removal of product typically cause reaction to proceed almost entirely one way. In a laboratory experiment, the reverse of reaction 2 could be made unimportant by measuring the initial rate of the reaction.

These equations have a close formal similarity to the Lindemann equations of Chapter 4, and the same sort of analysis will lead to a rate law. That is, the enzyme is typically present in small amounts, so that a steady-state concentration of ES will soon build up. We can therefore write

$$\text{rate of reaction} = V = k_2[\mathrm{ES}] \tag{5.37}$$

$$\frac{d[\mathrm{ES}]}{dt} = k_1[\mathrm{E}][\mathrm{S}] - k_{-1}[\mathrm{ES}] - k_2[\mathrm{ES}] = 0 \tag{5.38}$$

so that

$$[\mathrm{ES}] = \frac{k_1[\mathrm{S}]}{k_{-1} + k_2}[\mathrm{E}] \tag{5.39}$$

Now, if the experiment had been started by adding a certain amount of enzyme to the solution to produce a concentration $[\mathrm{E}]_0$, then

$$[\mathrm{E}] + [\mathrm{ES}] = [\mathrm{E}]_0$$

or

$$[\mathrm{E}] = [\mathrm{E}]_0 - [\mathrm{ES}] \tag{5.40}$$

at a later time. Substitution of this relationship for [E] into Eq. (5.39) leads to

$$[\mathrm{ES}] = \frac{k_1[\mathrm{S}][\mathrm{E}]_0}{k_{-1} + k_2 + k_1[\mathrm{S}]} \tag{5.41}$$

for the concentration of ES in the steady state. Substitution of this expression into Eq. (5.37) gives

$$V = \frac{k_1 k_2[\mathrm{S}][\mathrm{E}]_0}{k_{-1} + k_2 + k_1[\mathrm{S}]} = \frac{k_2[\mathrm{S}][\mathrm{E}]_0}{[\mathrm{S}] + (k_{-1} + k_2)/k_1} \tag{5.42}$$

The second term in the denominator is frequently written as K_m, the Michaelis constant. For many enzyme reactions k_2 is much less than k_{-1}, so that the steady-state concentration of ES is close to the equilibrium one for the first part of Eq. (5.36). In that case, K_m is close to the equilibrium constant for the dissociation of ES. We can write, then

$$V = \frac{k_2[S][E]_0}{[S] + K_m} \tag{5.43}$$

which is often called the Michaelis–Menten equation. This equation is in agreement with experimental observations that the rates of enzyme reactions are first order with respect to the total amount of enzyme present. Moreover, it correctly gives the limiting behavior at low and high substrate concentrations. At low values of [S],

$$V \rightarrow \frac{k_2[S][E]_0}{K_m} \tag{5.44}$$

which gives a first-order dependence on [S]. At high values of [S], this term rather than K_m dominates in the denominator to give

$$V \rightarrow k_2[E]_0 \tag{5.45}$$

That is, the rate of reaction no longer depends on the concentration of substrate. Physically, in this limiting case the enzyme has all the substrate molecules it can handle, the rate of reaction being limited by the rate at which the enzyme can process the available molecules. Essentially all the enzyme is in the ES form, and as soon as it dissociates to either E + S or E + P, another S molecule becomes associated with it.

Experimentalists usually prefer to use the Michaelis–Menten equation in one of several forms that leads to a straight-line graph. For example, simple inversion of Eq. (5.43) gives

$$\frac{1}{V} = \frac{[S] + K_m}{k_2[S][E]_0}$$

or

$$\frac{1}{V} = \frac{K_m}{k_2[E]_0}\frac{1}{[S]} + \frac{1}{k_2[E]_0} \tag{5.46}$$

so that a plot of $1/V$ versus $1/[S]$ will produce a straight line if the reaction is following this type of rate law. If $[E]_0$ is known, then values of k_2 and K_m may be found from the slope and intercept of the graph. Note that the y intercept $1/k_2[E]_0$ is the reciprocal of the limiting rate at high substrate concentrations, which by Eq. 5.45 is $k_2[E]_0$. Figure 5.2 (16) shows a typical set of data for an enzyme reaction, in both linear and direct plots.

A somewhat more realistic treatment than that of Eq. (5.36) would recognize that a catalytic reaction should have more symmetry with regard to reactants and products, so that the reaction path would probably look somewhat as shown in Fig. 5.3. The reaction scheme would then be

$$\mathrm{E + S} \underset{k_{-1}}{\overset{k_1}{\rightleftarrows}} \mathrm{ES}$$

$$\mathrm{ES} \xrightarrow{k_2} \mathrm{EP}$$

$$\mathrm{EP} \xrightarrow{k_3} \mathrm{E + P} \tag{5.47}$$

where EP represents the products, still associated with the enzyme, and we continue to consider reaction only in the forward direction. Steady-state treatment of these reactions leads to

$$V = \frac{k_2[\mathrm{S}][\mathrm{E}]_0}{[\mathrm{S}](1 + k_2/k_3) + (k_{-1} + k_2)/k_1} \tag{5.48}$$

This equation is identical to Eq. (5.42) except for the factor multiplying [S] in the denominator, which of course goes to 1 if the rate constant for

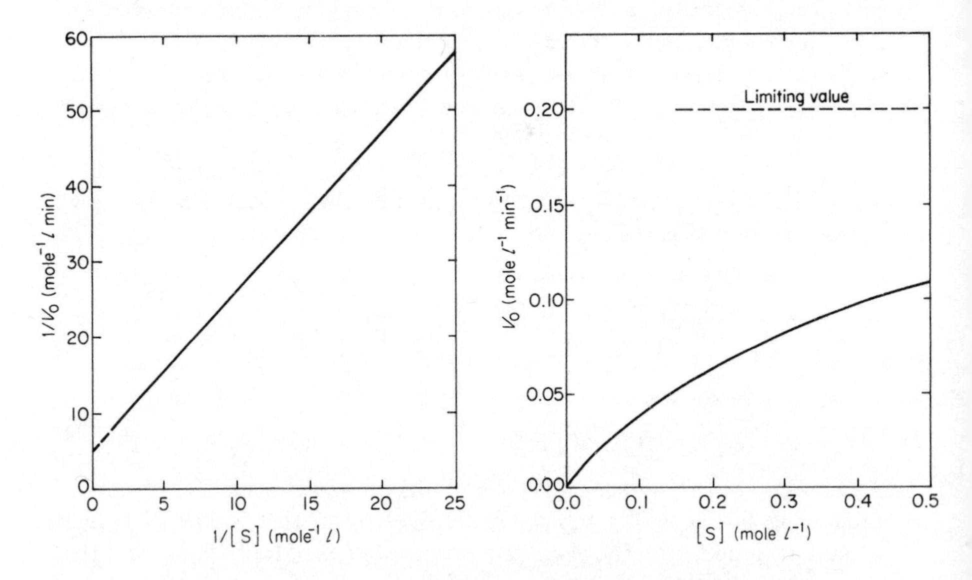

Fig. 5.2 Rates of hydration of acetaldehyde catalyzed by a fixed amount of bovine carbonic anhydrase at pH 7.09, 0°C, as a function of acetaldehyde concentration (16).

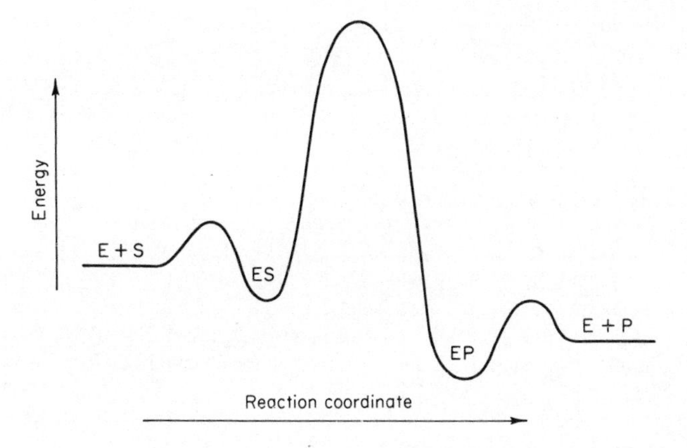

Fig. 5.3 Typical profile of potential energy along the reaction co-ordinate for an enzyme-catalyzed reaction. In general, products of the reaction may have more or less energy than reactants, and the activation energy for the ES → EP step is not always much larger than those for the other two steps.

the dissociation of EP is much greater than k_2, a fairly common occurrence as Figure 5.3 (which qualitatively reflects the behavior found for several enzymes) would suggest.

Rate constants for enzyme reactions vary in reasonably familiar ways with temperature, Arrhenius plots typically giving straight lines over moderate temperature ranges. Activation energies for association of substrate and enzyme tend to be no more than a few kilocalories per mole, while activation energies for conversion of ES to EP are usually less than 20 kcal mole^{-1}. For many reactions, enzymes provide lower-energy pathways than any other catalytic system. As an example (17) the decomposition of H_2O_2 in a pure water solution has a rate constant of about 10^{-7} mole^{-1} ℓ sec^{-1} and an activation energy of 18 kcal. In the presence of Fe^{2+} ions, the rate constant is 60 mole^{-1} ℓ sec^{-1}, with an activation energy of 10 kcal. With the enzyme catalase, the rate constant is 4×10^7 mole^1 ℓ sec^{-1}, with an activation energy of only 2 kcal. It is estimated that a single molecule of catalase can decompose 10^7 molecules of H_2O_2 per second at high substrate levels. It is not surprising, then, that enzyme kineticists have welcomed the relaxation methods mentioned in Section 5.5, and other methods for studying fast reactions described in Chapter 7, that have been developed in recent years. With them, it is often possible to measure rates of steps in the reaction that reach steady-state values too rapidly to be studied kinetically by the older methods.

References

1. H. Eyring and F. Daniels, *J. Amer. Chem. Soc.* **52,** 1473 (1930). See also H. Eyring and F. Daniels, *J. Amer. Chem. Soc.* **52,** 1486 (1930); F. Daniels and E. H. Johnston, *J. Amer. Chem. Soc.* **43,** 53 (1921); R. H. Lueck, *J. Amer. Chem. Soc.* **44,** 757 (1922).
2. R. B. Cundall, *Progr. React. Kinet.* **2,** 165 (1964).
3. E. N. Cain and R. K. Solly, *Int. J. Chem. Kinet.* **4,** 159 (1972).
4. R. H. Fowler and N. B. Slater, *Trans. Faraday Soc.* **34,** 81 (1938).
5. A. R. Fairclough and C. N. Hinshelwood, *J. Chem. Soc.* **1937,** 538.
6. L. P. Hammett, *Chem. Rev.* **17,** 125 (1935).
7. G. N. Burkhardt, W. G. K. Ford, and E. Singleton, *J. Chem. Soc.* **1936,** 17.
8. L. P. Hammett, "Physical Organic Chemistry." McGraw-Hill, New York, 1940; 2d Ed., McGraw-Hill, New York, 1970.
9. R. W. Taft, Jr., *J. Amer. Chem. Soc.* **74,** 2729, 3120 (1952); **75,** 4231 (1953).
10. R. W. Taft and I. C. Lewis, *J. Amer. Chem. Soc.* **80,** 2436 (1957).
11. R. W. Taft, E. Price, I. R. Fox, I. C. Lewis, K. K. Anderson, and G. T. Davis, *J. Amer. Chem. Soc.* **85,** 709 (1963).
12. P. Debye, *Trans. Electrochem. Soc.* **82,** 265 (1942).
13. V. Henri, *C. R. Acad. Sci.* **135,** 916 (1902).
14. L. Michaelis and M. L. Menten, *Biochem. Z.* **49,** 333 (1913).
15. G. E. Briggs and J. B. S. Haldane, *Biochem. J.* **19,** 338 (1925).
16. Y. Pocker and J. E. Meany, *Biochemistry* **4,** 2535 (1965).
17. K. J. Laidler, "The Chemical Kinetics of Enzyme Action." Clarendon Press, London and New York, 1958.

Further Reading

G. M. Fleck, "Chemical Reaction Mechanisms." Holt, New York, 1971 (a clear, elementary presentation with some good detailed examples).

A. A. Frost and R. G. Pearson, "Kinetics and Mechanism." 2d Ed. Wiley, New York, 1961 (a general text with emphasis on solution reactions).

L. P. Hammett, "Physical Organic Chemistry." McGraw-Hill, New York, 1940; 2d Ed., 1970 (two views, thirty years apart, of the continually developing field of organic reaction mechanisms).

J. E. Leffler and E. Grunwald, "Rates and Equilibria of Organic Reactions." Wiley, New York, 1963 (an effort to bring together everything that will help in understanding these properties.)

E. A. Moelwyn-Hughes, "The Kinetics of Reaction in Solution." 2d Ed., Clarendon Press, London and New York, 1947 (a thoughtful analysis of some problems that are still not all answered).

E. A. Moelwyn-Hughes, "Chemical Statics and Kinetics of Solutions." Academic Press, New York, 1971 (more quantitative study of solution kinetics).

P. D. Boyer, editor, "The Enzymes—Kinetics and Mechanism," Vol II, 3d Ed. Academic Press, New York, 1970.

K. J. Laidler, "The Chemical Kinetics of Enzyme Action." Clarendon Press, London and New York, 1958.

Problems

5.1 The volume of a mole of liquid benzene at room temperature is about 90 cc. What pressure would be required to compress a mole of an ideal gas to this volume at the same temperature? The answer may be thought of as an indication of the strength of the forces between molecules in liquids.

5.2 Glasstone, Laidler, and Eyring ("The Theory of Rate Processes," McGraw-Hill, New York, 1941) point out that diffusion may be thought of as a rate process governed by an equation

$$k_D = A_D \exp\left(-\frac{E_D}{RT}\right)$$

where E_D, the activation energy for diffusion, is the energy a molecule needs to force its way between neighboring molecules, and is typically a few kilocalories per mole, while A_D is the number of times per second the diffusing molecule would strike the walls of the little enclosure in which it resides. A_D for a molecule could be calculated from the average speed of a molecule and the average distance between collisions (a very short distance, let us assume 0.2 Å).

(a) Calculate the average speed of an acetone molecule at room temperature using the kinetic theory equation

$$\bar{c} = \left(\frac{8\mathbf{k}T}{\pi m}\right)^{1/2}$$

where m is the mass of the molecule in grams.

(b) Calculate the number of collisions per second experienced by a molecule.

(c) Calculate the number of collisions per second that would be experienced by an acetone molecule if it was in the gaseous phase at 1 atm and room temperature ($\sigma = 4.7$ Å).

(d) Calculate the number of times per second an acetone molecule changes its environment by diffusing to a new location, assuming an activation energy for diffusion of 5 kcal.

(e) If the acetone molecule moves 10 Å in each step of its diffusion (why would this be a reasonable value to use?), calculate the distance a molecule would move via the diffusion process per second. Compare this to the average speed $\bar{c}$.

5.3 We can find the basis for calculating the entropy of a single molecular vibration from the equation

$$G = H - TS$$

and the relation between the partition function for vibration Q_v and the free energy due to vibration,

$$G_v - H_0 = -RT \ln Q_v$$

which is analogous to Eq. (3.46). By rearranging the first equation we get

$$S = \frac{H - H_0}{T} - \frac{G - H_0}{T} = \frac{H - H_0}{T} + R \ln Q_v$$

As the vibrational frequency becomes small, $(H - H_0)/T$ approaches R while Q_v, which in general is equal to $(1 - e^{-h\nu/kT})^{-1}$, approaches $\mathbf{k}T/h\nu$ (Eqs. (3.54) and (3.57)).

(a) At 298° K, how low must the frequency ν be in order for the partition functions calculated by Eqs. (3.54) and (3.57) to differ by no more than 10%? Express this frequency in cm^{-1} units also.

(b) Assuming that with this low frequency $(H - H_0)/T$ can be taken to be R, calculate the entropy per mole associated with this vibration.

(c) Suppose the frequency of the activated complex was actually only one tenth of the value calculated in (a). Recalculate the entropy, and also calculate the entropy change on formation of 1 mole of the activated complex in the reaction of $C_2H_5Br + OH^-$, assuming reactant concentration of 1 mole/ℓ.

5.4 Jones and Phillips (*J. Chem. Soc.* **1971A,** 1881) reported the following data on the solution reaction

$$Cr(NH_3)_5F^{2+} + OH^- \rightarrow Cr(NH_3)OH^{2+} + F^-$$

with initial concentrations of 0.0005 M for the complex and 0.1 M for OH^- (from NaOH).

t (°C):	55.0	60.0	65.0	72.4	80.0
k (mole^{-1} ℓ sec^{-1} $\times 10^5$):	5.6	10.5	22.2	61.3	193.0

(a) Calculate what the rate constants ($k°$) would be at zero ionic strength. The following dielectric constant data for water may be useful:

t (°C):	20	40	60	80
ϵ:	80	74	68	60

(b) Calculate an Arrhenius equation for $k°$.
(c) Obtain values for $\Delta H_\ddagger$ and $\Delta S_\ddagger$. How much of the $\Delta S_\ddagger$ could reasonably be attributed to the charges on the reacting ions?

5.5 When rate data on groups of similar reactions are studied, it is frequently found that when the value of the Arrhenius A increases from one reaction to another, the value of E does also, so that the change in k is not as large as would be expected from the changes in either A or E individually. To some extent, this tendency is illustrated in the following data from a paper of Gay and Lalor (*J. Chem. Soc.* **1966,** 1179) on the reaction

$$Co(NH_3)_5X^{2+} + H_2O \rightarrow Co(NH_3)_5H_2O^{3+} + X^-$$

where X varied as shown in the tabulation below (temperature ~50° C):

X	A, (sec^{-1})	E, (kcal)
F	7.4×10^7	21.3
Cl	5.4×10^{11}	23.9
Br	2.5×10^{12}	23.8
NCS	1.1×10^{13}	30.7
NO_2	1.6×10^{14}	30.2
N_3	1.1×10^{16}	33.7

(a) Calculate the rate constant for each reaction at 50° C.
(b) Calculate ratios of the largest to smallest k, largest to smallest A, and $\exp(-E_S/RT)/\exp(-E_L/RT)$, where E_L and E_S are the largest and smallest activation energies.
(c) Plot a graph of $\log A$ versus E. From the best (even if not not very good) straight line, what is the slope of the graph? What would be the slope if the effects of A and E were to completely cancel one another?

5.6 (a) The logarithm of the acid dissociation constant of phenol in water at 25° C is −9.85. Using the data of Tables 5.3 and 5.4, calculate the dissociation constant of *m*-chlorophenol.
(b) If the rate constant for the base hydrolysis of *m*-hydroxy ethyl benzoate in 85% ethanol at 50° C is 0.0062 mole^{-1} ℓ sec^{-1}, what value would we expect for the rate constant for the hydrolysis of *p*-hydroxy ethyl benzoate under similar conditions?

5.7 Rates of reaction of substituted benzylamines with 1-chloro-2,4-dinitrobenzene (k') and with toluene-*p*-sulfonyl chloride (k'') were measured by Fischer, Hickford, Scott, and Vaughan (*J. Chem. Soc.*

1966, 466) at 45° C. Some of their data are given below:

Substituent	$k' \times 10^4 \pm 5\%$	$k'' \times 10^4 \pm 30\%$
H	18.8	14
m-Br	9.3	4
p-Br	11.6	4
m-Cl	10.1	4
p-Cl	11.0	4
m-CN	6.4	2
p-CN	5.9	0.5
m-CH_3O	17.3	30
p-CH_3O	25.4	30
m-CH_3	22.8	19
p-CH_3	23.3	40
m-NO_2	5.6	2
p-NO_2	5.4	—

From these data and the substituent constants of Table 5.4, calculate the two reaction constants.

5.8 Using the general approach of Section 5.5, derive an equation analogous to Eq. (5.33) that shows the approach to equilibrium for a first-order reaction A ⇄ B. Explain in words why the "rate constant" for this approach to equilibrium is larger than the forward rate constant for the reaction.

5.9 Suppose that the following data were found for an enzyme reaction at a constant enzyme concentration of 5×10^{-6} *M* with constant temperature, pH, and other relevant variables. Initial rates of substrate disappearance were made at several different initial concentrations of substrate, as follows:

Initial [S] (mole ℓ^{-1})	Initial rate (mole ℓ^{-1} sec^{-1})
.00060	.0031
.00066	.0034
.00084	.0042
.00121	.0052
.00190	.0069
.00274	.0092
.00645	.0115

From these data calculate the maximum reaction velocity that would be reached at very high substrate concentration, the Michaelis constant K_m, and the rate constant k_2, for conversion of reactant to product by the enzyme. If, in a different experiment using a relaxation technique, k_1 was found to be 1.5×10^7 mole^{-1} ℓ sec^{-1}, calculate the equilibrium constant for the dissociation of the enzyme–substrate complex, ES, and compare this to the Michaelis constant.

5.10 The rate of hydrolysis of 3′,5′-cyclic adenosine monophosphate by 3′,5′-cyclic nucleoside phosphodiesterase was studied photometrically by K. G. Nair (*Biochemistry* **5,** 150 (1966)). For a series of substrate concentrations at constant enzyme concentration, the following rates of reaction were measured on an arbitrary scale (that is, as change of optical density per unit time):

Monophosphate concentration (mole ℓ^{-1})	Rate
.00041	0.56
.00067	0.70
.00116	0.88
.00172	0.96
.00313	1.00

(a) From the slope and intercept of a plot of $1/V$ versus $1/[S]$, find the Michaelis constant K_m.

(b) Show from Eq. (5.46) that if this plot is extended to the left until the intercept with the x axis is found, this intercept will be $-K_m$. Make the extrapolation from your graph, and compare the value of K_m found this way with that from part (a).

5.11 Bonnichsen, Chance, and Theorell (*Acta. Chem. Scand.* **1,** 685 (1947)) measured the rate of decomposition of hydrogen peroxide catalyzed by the enzyme catalase. In the region in which the rate is first order with respect to both catalase and H_2O_2, they found that in an experiment with initial concentrations of 5.7×10^{-9} M catalase and 4×10^{-4} M H_2O_2, 90% of the H_2O_2 disappeared in 12 sec. Calculate the (second-order) rate constant for the reaction, and compare your answer with the value given in the text.

5.12 The authors referred to in Problem 5.11 state that the catalase–H_2O_2 reaction remains first order in H_2O_2, with no significant change in rate constant, up to an initial H_2O_2 concentration of at least 0.1 M. If a solution with concentrations 6×10^{-9} M in catalase and 0.1 M in H_2O_2 underwent 90% decomposition of H_2O_2 in 12 sec, calculate the number of molecules of H_2O_2 decomposed per catalase molecule in the first second of the experiment.

5.13 (a) In some cases it might happen (Bonnichsen, Chance, and Theorell considered it probably does for the catalase–H_2O_2

system) that the rate constant for decomposition of the Michaelis complex to products (k_2 in Eq. (5.42)) will be much greater than k_{-1} in the same equation. Show that, if this is so, the second-order rate constant described in Problem 5.11 is k_1, rather than the ratio k_2/K_m.

(b) Show that, if kinetic data are available only in the first-order region, the Michaelis constant cannot be found from graphs such as Fig. 5.2a. If it is not practical to go to higher substrate concentrations, what other approaches might be tried?

6 Reactions in Solids and Heterogeneous Systems

In the broad category of reactions in solids and heterogeneous systems fall many processes of great practical importance. For example, the production of high-strength metals and alloys, and the determination of the temperature ranges over which they can be used, involves a knowledge of rate processes. Typically, cooling a metal rapidly through a phase transition causes formation of small crystals of the low-temperature phase, and a relatively hard, perhaps brittle, metal. Heating to a moderate temperature will cause some crystal growth, leading to lower hardness but more toughness. Excessive heating may cause substantial crystal growth and so much loss of hardness that the metal would no longer be useful as a tool or a structural material. A considerable literature on the kinetics of phase changes and crystal growth in metals exists, but is difficult to appreciate without a more extensive background in metallurgy than most chemists have.

Similarly, there are very many biological processes that could best be described as heterogeneous since they involve movement of molecules through cell walls, or attachment of small molecules (such as oxygen or individual amino acids) to large ones (such as hemoglobin or growing proteins) for which some kinetic data are available, but again, because of the necessary background these reactions are best studied in biology or biochemistry courses.

To illustrate the principles involved in solid-state and heterogeneous reactions, let us look at two types of process for which the chemistry is relatively straightforward: the oxidation of metal surfaces, and the catalysis of gaseous reactions by solids.

6.1 OXIDATION OF METAL SURFACES

Nearly all metals are thermodynamically unstable toward oxidation by the air, but for many, a very thin oxide layer that forms rapidly on a newly created surface will act as a barrier to further oxidation, so that a very long time may elapse before the metal is actually oxidized. Typically, an oxide layer less than 100 Å thick, which is too thin to be observable by visible light, is sufficient to stop further reaction in metals such as aluminum, chromium, and nickel; hence these materials typically retain their characteristic metallic appearance for long periods, at least at room temperature. At higher temperatures, though, further oxidation can occur. Let us look at a typical example of oxidation at high temperature, that of cobalt metal.

Kinetic data have been obtained by two methods. In the first (1), samples of cobalt were exposed to an oxidizing atmosphere at constant temperature and oxygen pressure for varying lengths of time, following which they were cooled, cut in half perpendicular to the oxidizing surface, and examined microscopically, the thickness of the oxide product being measured, and any other interesting features noted. In the second method (2), a metal sample is suspended in an oxidizing atmosphere at constant temperature, and weight measurements made periodically. The latter method, as developed by E. A. Gulbransen, has produced the most extensive and accurate body of kinetic data. However, microscopic and x-ray and electron diffraction studies have frequently proved useful, and sometimes necessary, in determining the mechanism of the oxidation process, an essential step before the rates of oxidation can be understood in terms of chemical processes.

In the case of cobalt, microscopic examination established clearly the general mechanism of oxidation. Small platinum spots were placed on the surface of a cobalt sheet, which was then oxidized heavily. On microscopic examination, the platinum was found underneath the oxide layer on the surface of the remaining unoxidized cobalt. It was concluded that oxidation proceeded by diffusion of cobalt through the oxide that was already on the metal to the oxide–oxygen interface, and that the oxygen (or oxide ions) did not penetrate the metal at all.

X-ray diffraction and chemical analysis showed that, in the temperature range 1000–1350°C and for oxygen pressures of 1 atm and below, the oxide being formed was essentially CoO. (At higher oxygen pressures, Co_3O_4

Table 6.1 Metal/Oxygen Ratio in Cobalt Oxide[a]

Temperature (°C)	Co/O ratio	
	1 atm O_2	0.005 atm O_2
1000	0.994	0.998
1150	0.992	0.998
1350	0.990	0.998

[a] E. A. Gulbransen and K. Andrew, *J. Electrochem. Soc.* **98,** 241 (1951).

formed at the oxygen side, and CoO on the metal side, of the oxide layer.) However, the CoO was not exactly stoichiometric, the equilibrium oxide compositions present at various oxygen partial pressures being given in Table 6.1. In all cases, the number of metal ions is a little less than the number of oxygen ions, the defect clearly being greater in the high oxygen pressure experiments. Actually, this phenomenon is not unusual in metal oxides at high temperatures. In these compounds the oxide atoms (which have a radius of 1.40 Å) form a substantially complete crystal lattice, with very few atoms missing from the appropriate lattice points, while the metal atoms, which are much smaller (cobalt ions have a radius of 0.7 Å) fit into spaces between the oxide ions, and are much more able to move around in the lattice. Since the "CoO" must be electrically neutral, even if it varies from the stoichiometric ratio, a few of the ions must be Co^{3+}. The presence of a few vacancies, or holes, where cobalt ions would be expected to be, makes it easier for the other cobalt ions to move through the lattice. That is, a cobalt ion next to an empty space may move into that space with less energy than it would need to move into some other space that would not normally hold a cobalt ion. These x-ray and chemical experiments, then, are consistent with the reaction mechanism that was deduced from the platinum tracer experiments—metal ions diffusing through the oxide layer.

Further important information on the diffusion of cobalt ions in cobalt oxide is available (3). By a study of the movement of radioactive cobalt in cobalt oxides, the diffusion coefficients of cobalt ions were found to be:

$$\text{at } 1000°\text{C}, \ D = 2.6 \times 10^{-9}(P_{O_2})^{.35} \text{ cm}^2/\text{sec}$$

$$\text{at } 1150°\text{C}, \ D = 9.0 \times 10^{-9}(P_{O_2})^{.30} \text{ cm}^2/\text{sec}$$

$$\text{at } 1350°\text{C}, \ D = 5.1 \times 10^{-8}(P_{O_2})^{.28} \text{ cm}^2/\text{sec}$$

Clearly, as the oxygen pressure is lowered, the diffusion coefficient also drops, presumably because of the smaller number of cobalt vacancies at lower oxygen pressures, as shown in Table 6.1. By comparison of the data at various temperatures, an activation energy of 34.5 kcal is found—a value that, understandably, is much higher than the ~5 kcal found for diffusion in many liquids, as mentioned in Chapter 5.

The results of the actual oxidation experiments can be expressed in terms of the equation

$$W^2 = At \tag{6.1}$$

where W is the weight of oxide reacted per square centimeter of surface, A is a constant for a given temperature and pressure, and t is the time. Some values of the constant A are given in Table 6.2.

It should be noted that Eq. (6.1) does not apply from the very beginning of the experiment, and, indeed, we could hardly expect any simple relationship to apply. A real metal surface, although smooth-looking to the eye, contains many irregularities on a molecular scale. There will be crystal grain boundaries and many crystal lattice imperfections, at all of which the metal atoms will be in states of higher energy than those that are part of a "flat surface." Atoms with these above-average energies will react first, nucleating small crystals, which will grow until the surface is completely covered and oxygen gas can no longer come into direct contact with metal. Clearly, the detailed nature of the surface will affect the kinetics strongly during this initial period, but by the time an oxide layer of about 100 Å is built up, these surface details become unimportant, and the "parabolic law," so-called because of the algebraic form of Eq. (6.1), will prevail.

According to the above discussion, if the weight gain of a piece of metal is measured, and the square of the weight gain plotted against time, a straight line will be obtained after an initial period, and the slope of the straight line will give A. However, unless the metal already has an oxide

Table 6.2 Parabolic Rate Law Constants in the High-Temperature Oxidation of Cobalt[a]

Temperature (°C)	O_2 pressure	A (grams O_2 reacted/ cm^2 surface)^2
1150	1.0	9.3×10^{-8}
1150	0.15	5.0×10^{-8}
1150	0.0055	2.1×10^{-8}
1000	1.0	2.4×10^{-8}
1350	1.0	78×10^{-8}

[a] R. E. Carter and F. D. Richardson, *Trans. AIME* **203**, 336 (1955).

coating, this line will not go through the origin, and in order to use Eq. (6.1), one would choose as a new origin, at which t and W would be set to zero, a point on the curve at which straight-line behavior has been reached.

This parabolic rate law has been observed for the oxidation of many metals where the oxide film appears to be continuous and tightly adhering. It does not fit the oxidation data for active metals such as sodium, which forms a loose, crumbly oxide that allows penetration of oxygen gas.

It was Wagner (4) who outlined and then put into quantitative form the basis for the parabolic law, pointing out the electrochemical nature of the process, and the fact that the oxide layer already on the surface of the metal is in a steady-state condition, but a dynamic one. That is, oxygen molecules absorbed on the surface of the oxide will tend to become oxide ions, by picking up electrons from the Co^{2+} ions, converting them to Co^{3+}. Electrons will flow through the oxide layer (probably via the $Co^{2+} \rightleftarrows Co^{3+}$ reaction) until the outside of the oxide layer becomes substantially negative with respect to the inside. The steady-state voltage across the oxide layer will correspond to that for the oxidation reaction based on the equation

$$\Delta G = -n\mathfrak{F}E \tag{6.2}$$

where ΔG is the free energy of reaction, $\mathfrak{F}$ the Faraday, n the number of electrons involved in the reaction, and E the voltage. This voltage, along with the variation in cobalt ion concentration from one side of the film to the other, will provide the driving force for the flow of metal ions from the metal–oxide surface to the oxygen–oxide surface. As the oxide layer grows, it is clear that the same potential and concentration differences across the layer will continue, so that the driving force for reaction will be inversely proportional to the thickness of the layer. This would lead us to write

$$\frac{dW}{dt} = \frac{C}{W} \tag{6.3}$$

where C is a constant, and W and t have the same meaning as before. If we set t and W to zero at a point at which the oxidation process has advanced far enough so the parabolic law is obeyed, then

$$\int_0^W W\,dW = \int_0^t C\,dt$$

$$\frac{W^2}{2} = Ct \qquad \text{or} \qquad W^2 = At$$

which is the rate law found experimentally.

Wagner's analysis, though, goes far beyond this qualitative agreement with experiment. His quantitative treatment starts with an equation similar to (6.3), written as

$$\frac{dn_{\text{equiv}}}{dt} = \frac{k_r}{\Delta x} \tag{6.4}$$

where n_{equiv} is the number of equivalents of oxygen in the oxide layer per square centimeter of surface, when the oxide layer is 1 cm thick; Δx is the thickness of the oxide layer at a given time, in centimeters, and k_r is called the "rational rate constant" and can be thought of as the number of equivalents of oxygen per second that will be added to the oxide layer on 1 cm² of surface, when the oxide layer is 1 cm thick.

Let us first note the relationships between n_{equiv}, Δx, and W. Since oxide ions have a double negative charge, we would write

$$n_{\text{equiv}} = \frac{2W}{M_{\text{O}}} \tag{6.5}$$

where M_{O} is the atomic weight of oxygen. The weight of oxygen in the oxide layer on a square centimeter of surface would be

$$W = \rho\Delta x \frac{M_{\text{O}}}{M_{\text{CoO}}} \tag{6.6}$$

where ρ is the density of the oxide, so that

$$\Delta x = \frac{WM_{\text{CoO}}}{\rho M_{\text{O}}} \tag{6.7}$$

If we substitute these values into Eq. (6.4) and integrate, we get

$$\frac{2\,dW}{M_{\text{O}}\,dt} = \frac{k_r}{WM_{\text{CoO}}/\rho M_{\text{O}}} \tag{6.8}$$

$$\frac{2M_{\text{CoO}}}{\rho M_{\text{O}}^2} \int_0^W W\,dW = k_r \int_0^t dt$$

$$\frac{M_{\text{CoO}}W^2}{\rho M_{\text{O}}^2} = k_r t$$

$$W^2 = \frac{\rho M_{\text{O}}^2}{M_{\text{CoO}}} k_r t \tag{6.9}$$

Accordingly, we can identify

$$A = \frac{\rho M_{O}{}^{2}}{M_{CoO}} k_r \tag{6.10}$$

Let us now go on to calculate k_r, using the idea that the driving force for diffusion of the cobalt ions through the oxide is primarily an electrostatic one. To make the idea clear, let us assume at first that the diffusion coefficient is independent of oxide composition, although in fact this is not true, and will be allowed for later. For diffusion of charged particles, the Nernst–Einstein equation states that

$$v = \frac{FD}{\mathbf{k}T} \tag{6.11}$$

where v is the velocity of the particle (cm/sec), F is the force (dynes) on the particle, D is the diffusion constant (cm^2 sec^{-1}), $\mathbf{k}$ is the Boltzmann constant, and T is the temperature. The force on a charged particle is neE_e, where n is the number of electronic charges on the particle (2 for most of our cobalt ions), e is the electronic charge (esu), E_e is the field strength (esu/cm) which is 1/300 of the field strength (V/cm).

Finally, the field strength (V/cm), considering that k_r refers to a layer of oxide 1 cm thick, is equal to the reaction potential given by Eq. (6.2), where ΔG is the free energy change for the reaction

$$Co(s) + \tfrac{1}{2} O_2(g) \rightarrow CoO(s)$$

under the experimental oxygen pressure, measured in joules. Accordingly,

$$E = \frac{\Delta G}{n\mathfrak{F}}$$

$$E_e = \frac{\Delta G}{300 n\mathfrak{F}}$$

$$F = \frac{e\Delta G}{300\mathfrak{F}}$$

$$v = \frac{e\, \Delta G\, D}{300\mathfrak{F}\mathbf{k}T} \tag{6.12}$$

The value of k_r (the number of equivalents passing a given plane in the oxide layer per second) will be

$$k_r = \frac{2\rho v}{M_{CoO}} = \frac{2\rho e\, \Delta G\, D}{300\mathfrak{F}\mathbf{k}TM_{CoO}} \tag{6.13}$$

and

$$A = \frac{2\rho^2 e \,\Delta G\, D}{300\mathfrak{F}\mathbf{k}T} \frac{M_{O}^2}{M_{CoO}^2} \tag{6.14}$$

For 1150°C and 1 atm oxygen pressure, the numerical values are approximately

$$\Delta G = 2 \times 10^5 \quad \text{J}$$

$$D = 9 \times 10^{-9} \quad \text{cm}^2/\text{sec}$$

$$\rho = 6 \quad \text{g/cc}$$

so

$$A = \frac{2 \times 6 \times 6 \times 4.8 \times 10^{-10} \times 2 \times 10^5 \times 9 \times 10^{-9} \times 16 \times 16}{3 \times 10^2 \times 9.65 \times 10^4 \times 1.38 \times 10^{-16} \times 1.423 \times 10^3 \times 75 \times 75}$$

$$= 5 \times 10^{-7} \quad (\text{g cm}^2)^2 \text{ sec}^{-1}$$

Comparison of this quantity with the data of Table 6.2 shows that the calculated value is on the high side, which is due to the assumption that the diffusion coefficient of cobalt ions measured in CoO at 1 atm applies at the lower effective oxygen pressures inside the oxide layer. Since Carter and Richardson found a significant dependence of the diffusion coefficient on the oxygen pressure, the product $\Delta G\, D$ in Eqs. (6.13) and (6.14) should be replaced by an integral

$$\int D \, d\Delta G$$

where the free energy would vary from that at the metal–oxide interface to that at the oxide–oxygen interface. Since the diffusion coefficient is given in terms of oxygen pressure, it is convenient to express the free energy in this unit also, using the expression

$$\Delta G = RT \ln \frac{P_2}{P_1}$$

or

$$d\Delta G = RT\, d \ln P = RT \frac{dP}{P}$$

In our example, the upper limit of pressure is 1 atm, while the lower limit is found from the free energy change for the oxidation reaction, about 2 ×

10^5 J, so that, on a mole basis, we may write

$$\ln P = \frac{2 \times 10^5}{8.3 \times 1423} = -17.0$$

from which $\log P = -7.37$, and $P = 4 \times 10^{-8}$ atm. This is the dissociation pressure of O_2 over CoO when it is in equilibrium with metallic cobalt, the situation existing at the metal–oxide interface.

Accordingly, using the diffusion constant equation given above, the integral becomes

$$\int D\, d\Delta G = \int_{4\times10^{-8}}^{1} 9 \times 10^{-9} P^{0.3} (8.314 \times 1423) \frac{dP}{P}$$

$$= 1.065 \times 10^{-4} \int_{4\times10^{-8}}^{1} P^{-0.7}\, dP$$

$$= \frac{1.065 \times 10^{-4}}{0.3} [(1)^{0.3} - (4 \times 10^{-8})^{0.3}]$$

$$= 3.5 \times 10^{-4} \quad (\text{J cm}^2\ \text{sec}^{-1})$$

Replacement of the produce $\Delta G\, D$ in Eq. (6.14) (which in our example had the value $2 \times 10^5 \times 9 \times 10^{-9} = 18 \times 10^{-4}$) by the integral leads to the value 1.1×10^{-7} $(\text{g/cm}^2)^2\ \text{sec}^{-1}$ for the weight gain constant, in quite good agreement with the experimental value of 9.3×10^{-8}, from Table 6.2.

In this analysis, it has been assumed that the electrical forces predominate in causing diffusion. We may show the correctness of this assumption by calculating the rate of diffusion that would occur if caused only by the concentration gradient. Fick's first law of diffusion will apply here:

$$J = -D \frac{dc}{dx} \tag{6.15}$$

where J is the flux, or the rate at which material crosses a surface perpendicular to a concentration gradient, and c is the concentration of the diffusing species. J and c can be in any consistent units—we could use equivalents/cc to fit Wagner's rational rate constant. The minus sign in the equation simply allows for the fact that diffusion occurs from high to low concentrations. Across a 1-cm layer of oxide at 1150°C the concentration gradient will be

$$\frac{dc}{dx} = \frac{2\rho}{M_{\text{CoO}}} (S_{\text{O}} - S_{\text{M}}) \tag{6.16}$$

where the term in parentheses includes the stoichiometric coefficients for Co_sO at the oxide–oxygen interface ($S_O = 0.992$ at our experimental conditions, from Table 6.1) and at the metal–oxide interface (S_M is probably very close to 1.000). We can therefore write

$$k_r = J = \frac{2\rho}{M_{CoO}} (S_M - S_O)D \tag{6.17}$$

Combination of this expression with Eq. (6.10) indicates that the contribution to A of the concentration-induced diffusion is

$$A_D = \frac{2\rho^2 M_O{}^2}{M_{CoO}{}^2} (S_M - S_0)D \tag{6.18}$$

Even using the largest value of D, the one applying at 1 atm oxygen pressure, the numerical value of this quantity is only 2.4×10^{-10} $(g/cm^2)^2$ sec^{-1}, a very small fraction of that due to the electrostatic driving force.

The Wagner theory, then, is a very successful one that quantitatively correlates rates of oxidation with related chemical and physical quantities in the reacting system. In one sense, the rate of oxidation is diffusion controlled, but the rate of diffusion is strongly influenced by the chemical potential (free energy change) of the reacting system.

It is probably correct to say that the above derivation would have been simpler without the several different sets of units that were used. It would probably be a good idea to redefine an SI system rational rate constant in terms of the number of moles of oxygen added to the oxide layer per square meter of surface per second, for a thickness of 1 meter. Such an exercise is outlined in Problem 6.1.

The oxidation of titanium and zirconium metals involves an interestingly different process-diffusion of oxygen into the metal itself, with the formation of metal–oxygen "alloys" that retain the crystal structure of the metal, while containing continuously varying amounts of oxygen. Initial reaction of oxygen with a hot titanium surface leads to the formation of a film of oxide, which grows thicker with time by the same general process followed by cobalt. Here, the chemistry in the oxide layers is more complicated because a series of oxides of titanium exists, ranging from TiO_2 at the oxide–oxygen interface, to TiO at the metal–oxide interface, and at the same time the variations of these compounds from the stoichiometric compositions is greater than that found for CoO.

The point of special interest, however, is the diffusion of oxygen into the metal, the rate of which is governed by Fick's second law of diffusion

$$\frac{dc}{dt} = D \frac{d^2c}{dx^2} \tag{6.19}$$

This equation indicates that if we examine a certain part of the metal at several times, t, the concentration of oxygen will change with time. Oxygen will diffuse through the part we are looking at, but more will diffuse in than out, the net change being given by the equation. The oxygen concentration will continue to rise as long as a supply is available at the edge of the titanium metal surface, and as long as the solution of oxygen in titanium remains unsaturated.

Some complications should be considered. As with the diffusion of cobalt ions in CoO, the diffusion constant D will be a function of concentration. It seems fairly clear, however, that in titanium the oxygen occupies positions between the titanium atoms (interstitial positions) and since there are many more such positions than there are oxygen atoms, D should not be expected to depend as strongly on composition, or equilibrium oxygen pressure, as in the cobalt example. Accordingly, we will assume a constant D in our calculations, although the problem can be solved with a concentration-dependent D if more-complicated mathematics are used. Second, if titanium atoms are diffusing through the oxide layer at the same time that oxygen is diffusing into the metal, then the condition of a fixed medium in which diffusion is occurring (a basic assumption of the Fick equation) is not fulfilled. Again, for simplicity, let us assume that the experimental conditions are such that the loss of titanium from the metal is negligible under our experimental conditions. This condition could be fulfilled, in principle, by exposing a titanium metal surface to a low oxygen pressure, equal to the dissociation pressure of TiO in equilibrium with a saturated solution of oxygen in titanium. Finally, titanium metal has a hexagonal crystal structure under typical oxidizing conditions, and the value of D probably is different in the two principal crystal directions. Actually, most samples of metals, including titanium, consist of many small crystals more or less randomly oriented, and diffusion constants reported for oxygen in titanium have been for polycrystalline samples, so it seems reasonable to assume no directional variation of D.

Because of all these factors that could complicate results, it is not too practical to try to reproduce experimental oxidation data, as was done for cobalt. However, we can use Fick's second law of diffusion to calculate the development of the titanium–oxygen solid solution that would occur under the idealized assumptions given above. Given these assumptions, integration of Eq. (6.19) yields

$$c(x, t) = C_{\mathrm{O}}\left(1 - \operatorname{erf}\left(\frac{x}{2Dt}\right)\right) \tag{6.20}$$

where C_{O} is the concentration of oxygen atoms in the metal at the metal–oxide interface, and the other variables have already been defined. The

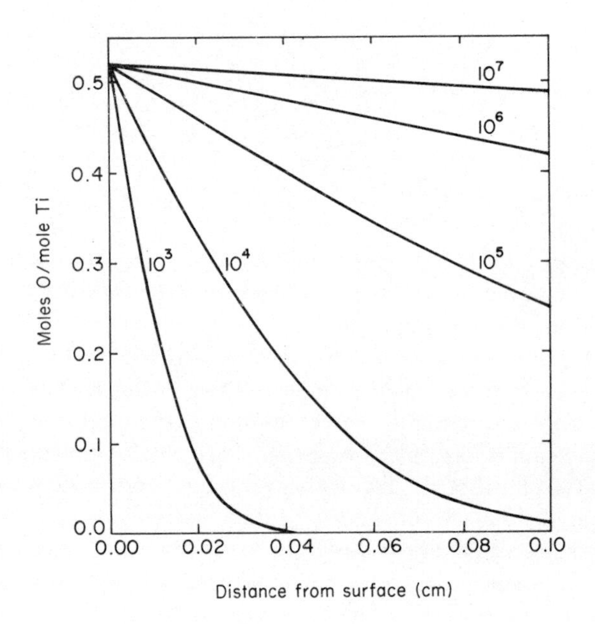

Fig. 6.1 Profiles of oxygen concentrations in a thick sheet of titanium, calculated at five different times by Eq. (6.20), assuming $D = 10^{-7}$ cm^2 sec^{-1} and $C_O = 0.52$ atoms of O per atom of titanium. Times are given in seconds (10^7 sec $\approx$ 4 months).

symbol "erf" stands for the "error function," a quantity available in standard mathematical tables. As the quantity $x/(2Dt)$ varies from 0 to a large number, the error function increases from 0 to 1. One other small assumption in the integration is that the piece of metal is thick compared to the layer affected by the oxygen, so that it does not become saturated. In Fig. 6.1 a series of concentration profiles are shown, at different times, for diffusion of oxygen into titanium at a temperature of 800°C, at which D has been found to be about 10^{-7} cm^2/sec (5). The saturated solution of oxygen in titanium at the metal–oxide interface contains 34 at. % O, or 0.52 atoms O per atom Ti, and this value has been taken as C_o. Since the titanium lattice size changes relatively little as the oxygen is added, this is as good a way of expressing concentration as any other.

The oxidation of iron in dry air or oxygen follows a course similar to that described for cobalt, but the process tends to be more rapid since the variable oxidation state of iron leads to oxides that have relatively large numbers of vacancies in their crystal lattices, and therefore relatively large diffusion coefficients. Even so, iron oxidizes very slowly at room temperature in dry air.

Rusting of iron in the presence of water is considered to involve a different sort of electrochemical process which is so complicated that no real kinetic description has yet been formulated, although extensive bodies of empirical data are available, and the understanding of the qualitative mechanism of the process is sufficient to suggest many ways of reducing rust formation.

If a piece of iron or steel is examined microscopically, its surface will be seen to be quite heterogeneous. The carbon content varies slightly from place to place, as do the concentrations of other alloying elements. If the iron is immersed in water, small electrochemical cells will be set up among these various areas. For simplicity, we may concentrate our attention on two such areas, the first being an unusually pure iron area, and the second one that contains more carbon than usual. The tendency will be for iron to go into solution from the first area

$$Fe \rightarrow Fe^{+2} + 2e$$

and for the iron in solution to go toward the second area, but the chance that it will actually deposit there is small since iron has less tendency to deposit at a cathode than other substances that can be in solution, such as hydrogen ions. The most likely reaction to occur at the second area will be

$$2e + 2\,H^+ \rightarrow H_2$$

If the water in contact with the iron is very pure and free of dissolved gases, this reaction will not proceed far. The current flow will be very small, and the build-up of tiny bubbles of H_2 will slow down the cathode reaction. In a more typical situation, though, there would be dissolved electrolytes (at least CO_2 from the air, which would give H^+ ions) and also dissolved oxygen. The oxygen would react with the hydrogen, removing it from the cathodic areas, or else react directly with the hydrogen ions

$$4\,H^+ + 4e + O_2 \rightarrow 2\,H_2O$$

near the cathode surface, thus ensuring that this area was always free for the conduction of electricity. Moreover, the O_2 will react with the Fe^{2+} in solution, producing most often a rust consisting of hydrated Fe_2O_3:

$$4\,Fe^{+2} + O_2 + 4\,H_2O \rightarrow 2\,Fe_2O_3 + 8\,H^+$$

Typically, then, in the first stages of iron rusting, the purer parts dissolve away first, so the corroded surface becomes pitted and uneven. Rust will tend to form most rapidly wherever the access of oxygen is greatest. For example, if an iron surface is partially immersed in water, rust will form near the surface of the water, where the oxygen concentration is highest, while iron loss from the surface will be greatest at greater depths, where the oxygen concentration is less. Similarly, if a piece of rust is adhering to a certain area, the oxygen concentratoin under the spot will be less. Iron will tend to dissolve under the area of rust, while rust will continue to appear around the periphery of the area.

While chemical ideas are helpful in leading to a general understanding of the process of rusting, quantitative analysis that would allow for the multitude of variables present in practical situations has not so far been possible.

6.2 CATALYSIS OF REACTIONS BY SOLIDS

In this section we will return to more familiar ground since the reactions to be considered will be gaseous ones similar to some we have looked at earlier. The first example will be one that was first studied in depth by Langmuir over fifty years ago: the oxidation of carbon monoxide, catalyzed by a platinum surface (6). This process is of current interest as a method for removing CO from automobile exhaust gases.

The carbon monoxide reaction in the gas phase is extremely slow compared, for example, to the hydrogen–oxygen reaction. Shock-tube studies of comparable gas mixtures in shock tubes (7) have shown that to obtain given "induction times," temperatures of CO–O_2 mixtures must be several hundred degrees hotter than corresponding H_2–O_2 mixtures. This is not surprising when one considers that both CO and O_2 are multiply bonded molecules, with dissociation energies of about 250 and 120 kcal, whereas H_2 has a single bond with the lower dissociation energy of 100 kcal. In the presence of a clean platinum surface, appreciable reaction between CO and O_2 has been observed at temperatures as low as 100°C (8), at which temperature the gas mixture would remain unreacted for geological times in the absence of a catalyst.

Experiments have been carried out over a series of low pressures from 10^{-7} to 10^{-3} atm, temperatures from 350 to 1500°K, and at various CO/O_2 mole ratios. Over this range of conditions a variety of behaviors has been observed.

(a) At moderate temperatures and pressures, and at low pressures when the temperature is also low, the speed of the reaction is directly proportional to the O_2 pressure but inversely proportional to the CO pressure. Incidentally, these pressure effects lead to an overall zero-order reaction—the rate of reaction is independent of the total gas pressure. This is one of the relatively few examples of this reaction order.

(b) At high temperatures and high CO/O_2 ratios, the rate of reaction is governed by the rate at which O_2 molecules strike the platinum surface (the last quantity can be calculated by the kinetic theory).
(c) At high temperatures and low CO/O_2 ratios, the rate of reaction is governed by the rate at which CO molecules strike the platinum surface.
(d) In the region of moderate temperatures and pressures, the rate of reaction for a given pressure and composition is expressible as an Arrhenius equation, with an activation energy equal to that for the rate of evaporation of a CO film from a platinum surface.
(e) At high temperatures and low pressures, the rate of reaction is nearly independent of temperature.

Along with these limiting conditions, there are many observations in intermediate regions of pressure, temperature, and composition in which the variations of reaction rate followed more-complex rate laws.

Langmuir was able to explain his observations in terms of a relatively simple mechanism. The following discussion is based essentially on his analysis, with a few changes, partly for simplification and partly because of insights from more recent work. Suppose, then, that we make these assumptions:

(1) The platinum surface is essentially uniform, or at least acts that way, down to a molecular scale, so we visualize a clean platinum surface as a smooth crystal plane, on which the gases may be adsorbed.
(2) Adsorbed gases form only a monomolecular layer on the platinum surface. Once a platinum atom is covered by an adsorbed molecule, another molecule will not be adsorbed on top of the first. If a molecule lands on an unoccupied platinum atom, it will always be adsorbed.
(3) Carbon monoxide molecules are adsorbed as such on the platinum surface. They are strongly adsorbed (in other studies, Langmiur found the activation energy for loss of CO from a platinum surface to be 32.4 kcal) but apparently do not react with the surface. Moreover, the CO molecules on the surface are unreactive toward oxygen in either the gaseous or adsorbed states. Finally, a steady state exists between CO molecules on the surface and those in the gas phase. CO molecules are continually landing on unoccupied platinum atoms, while other CO molecules are going from the adsorbed state to the gas.
(4) Oxygen molecules are not adsorbed as such on the platinum surface. If an oxygen molecule strikes an unoccupied platinum atom, it dissociates very quickly, one atom remaining on the first platinum atom, and one going to a nearby unoccupied site. Oxygen atoms are held very tightly by the platinum, and rarely recombine and return to the gas phase.
(5) Reaction occurs chiefly when a CO molecule from the gas phase

strikes an adsorbed oxygen atom. Formation of CO_2 (like adsorption) occurs with a negligibly small activation energy, and the CO_2 molecules formed leave the surface immediately. Moreover, most experiments were carried out with a trap to capture CO_2 molecules, so that CO_2 molecules, once formed, effectively disappeared from further consideration.

These qualitative ideas may be put into a quantitative form, using the following quantities:

θ_1 and θ_2 are the *fractions* of surface covered by CO molecules and O atoms, respectively. We will assume that each of these species occupies one platinum atom. θ (unsubscripted) is the fraction of surface free of adsorbed gases.

μ_1 and μ_2 are the number of moles of CO and O_2 molecules striking 1 cm^2 of surface per second. From the kinetic theory,

$$\mu_1 = \frac{p}{(2\pi MRT)^{1/2}} \tag{6.21}$$

where p is the pressure in dyn/cm^2 and R is in erg $mole^{-1}$ deg^{-1}.

ν_1 is the number of moles of CO that would leave 1 cm^2 of completely covered surface per second. We will not use a quantity ν_2 since it is assumed that all O atoms remain on the surface until they react.

W is the rate of the reaction

$$2\ CO + O_2 \rightarrow 2\ CO_2$$

on 1 cm^2 of surface. The rate will be expressed in mole sec^{-1} for 1 cm^2 of surface, and, in line with the definitions of Chapter 2, the rate of formation of CO_2 will be $2W$ since its stoichiometric coefficient is two.

Using these variables, and the five postulates given above, we may write that, for the adsorption and desorption of CO,

$$\mu_1\theta = \nu_1\theta_1 \tag{6.22}$$

For the adsorption of oxygen and removal of oxygen by reaction with CO, a steady state will result if

$$2\theta\mu_2 = \mu_1\theta_2 \tag{6.23}$$

For the rate of reaction in terms of CO striking oxygen atoms,

$$2W = \mu_1\theta_2 \tag{6.24}$$

Since the surface must be either covered or uncovered,

$$\theta + \theta_1 + \theta_2 = 1 \tag{6.25}$$

A general equation may be found by solving for each of the θ's, and substituting into Eq. (6.25). From Eq. (6.24),

$$\theta_2 = \frac{2W}{\mu_1}$$

Combination with (6.23) gives

$$\theta = \frac{W}{\mu_2}$$

while further combination with (6.22) gives

$$\theta_1 = \frac{W\mu_1}{\nu_1\mu_2}$$

$$\frac{W}{\mu_2} + \frac{W\mu_1}{\nu_1\mu_2} + \frac{2W}{\mu_1} = 1 \tag{6.26}$$

or

$$W = \left(\frac{1}{\mu_2} + \frac{\mu_1}{\nu_1\mu_2} + \frac{1}{\mu_1}\right)^{-1} \tag{6.27}$$

Limiting cases will be expected to arise when one of the terms in the denominator dominates over the others, to give a simple expression. These will correspond to any of θ, θ_1, or θ_2 approaching 1. Let us look at the five limiting cases in this way.

(a) At moderate temperatures and pressures, the surface is largely covered by CO molecules, so $\theta_1 \to 1$, and

$$W \to \frac{1}{\mu_1/\nu_1\mu_2} = \nu_1 \frac{\mu_2}{\mu_1} \tag{6.28}$$

Since $\mu_2 \mathrel{\dot{\approx}} p_2 \mathrel{\dot{\approx}} [O_2]$ and $\mu_1 \mathrel{\dot{\approx}} p_1 \mathrel{\dot{\approx}} [CO]$, this gives the reaction orders found with respect to O_2 and CO, and the overall zero order.

(b) At high temperatures and high CO/O_2 ratios, the CO concentration on the surface will be low because of the high temperature, and the O atom concentration will be low because of the small O_2 gas concentration and the fast rate of removal of O atoms by CO collisions, so $\theta \to 1$, and

$$W \to \mu_2 \tag{6.29}$$

This is a mathematical statement of the experimental behavior.

(c) At high temperatures and low CO/O_2 ratios, the surface will become covered with O atoms, that is, $\theta_2 \to 1$, and

$$W \to \mu_1 \tag{6.30}$$

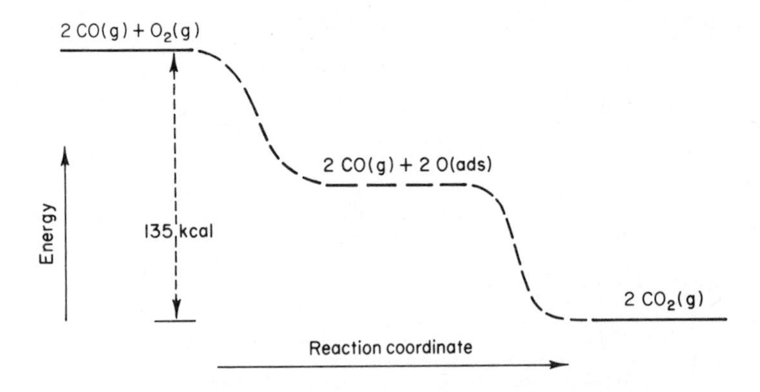

Fig. 6.2 Hypothetical potential energy profile for the platinum-catalyzed oxidation of CO. The enthalpy change is 135 kcal at 500°K.

Physically, every CO molecule that strikes the surface will encounter an O atom, as apparently found experimentally.

(d) The activation energy at moderate temperatures and compositions, as given by Eq. (6.28), is that for ν_1, the rate of loss of CO from the platinum surface.

(e) At high temperatures, the rate of reaction depends only on numbers of collisions, which have very small dependence on temperature.

The model, then, gives an excellent interpretation of the general nature of the catalytic process, with quantitative prediction of reaction rates (based on the kinetic theory) in some instances. The experimental observations indicate that neither adsorption of oxygen nor reaction of CO with adsorbed oxygen involves a significant activation energy. Since the reaction

$$2\,CO + O_2 \rightarrow 2\,CO_2$$

is substantially exothermic, one would conclude that the energy profile of the reaction would be somewhat as shown in Fig. 6.2.

Without doubt, the model presented above is simplified from reality. For example, it is known that a small fraction of O_2 molecules will react directly, on collision, with adsorbed CO molecules, and an allowance could be made for this effect. More importantly, though, a skeptical chemist might ask if there is any other mechanism that could explain the data; and the answer to this question is yes! If it is assumed that molecules of O_2 do not dissociate on the surface, but remain as adsorbed O_2 until struck by a CO molecule, which removes one O atom and then is desorbed, leaving an O atom on the surface, and if the rest of the Langmuir mechanism is left as is, application of the same reasoning process leads again to Eq. (6.26), which explains all the observations. These two mechanisms, then, are indistinguishable by the experiments that were done, and there could well be other reasonable mechanisms that are consistent with observation. At present, as indicated in recent papers, the mechanism of this catalytic process is still not known for sure.

The rearrangement of normal butenes on oxide catalysts is an example taken from the many hydrocarbon rearrangements that are carried out in the petroleum and petrochemical industries. Some other processes, such as the rearrangement of straight chain to branched hydrocarbons to improve performance in internal combustion engines, and "cracking," to convert higher to lower molecular weight hydrocarbons, are carried out industrially on a much larger scale, but they are too complex to be very good textbook examples.

Let us start by looking at these hydrocarbons and their interrelationships. Some important thermochemical properties are given in Table 6.3. The data indicate that at 350°K, which lies in the middle of the temperature range where experimental data have been obtained, *trans*-2-butene is the most stable of the three compounds, but the difference in free energies of formation is small enough that all forms will be present in significant amounts at equilibrium. To calculate the equilibrium situation, consider

Table 6.3 Some Thermodynamic Properties of the Normal Butenes[a]

Compound	Formula	Enthalpy of formation, (kcal/mole, 350°K)	Free energy of formation (kcal/mole, 350°K)
1-butene	$CH_2{=}CH\diagdown CH_2 \diagup CH_3$	−0.8	20.2
cis-2-butene	$CH_3\diagdown CH{=}CH \diagup CH_3$	−2.5	18.9
trans-2-butene	$CH_3\diagdown CH{=}CH \diagdown CH_3$	−3.4	18.2

[a] Data from API Tables, see Ref. 12.7, Chapter 2.

all three forms at once:

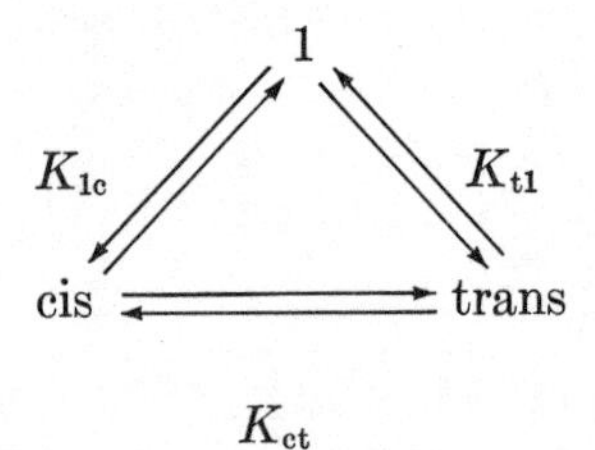

From the general equation for an equilibrium constant,

$$\Delta G^\circ = RT \ln K$$

we can write

1-butene $\rightleftarrows$ *cis*-2-butene, $\Delta G^\circ = 18.9\text{–}20.2$ kcal; $K_{1c} = 7.1$
cis-2-butene $\rightleftarrows$ *trans*-2-butene, $\Delta G^\circ = 18.2\text{–}18.9$ kcal; $K_{ct} = 2.6$
trans-2-butene $\rightleftarrows$ 1-butene, $\Delta G^\circ = 20.2\text{–}18.2$ kcal; $K_{t1} = 0.054$

Each of these equilibrium constants can be thought of as a ratio of either pressures or concentrations since the gases will behave reasonably ideally. For example,

$$K_{1c} = \frac{p_c}{p_1} = \frac{[\text{cis-2-butene}]}{[\text{1-butene}]} \tag{6.31}$$

It is easily seen that the ratio of pressures of the three gases does not depend on the total pressure, and is approximately $p_t{:}p_c{:}p_1 = 0.70{:}0.26{:}0.04$ at equilibrium.

There is an additional thermodynamic relationship. Since the three equilibria exist in a cycle, the equilibrium constant between 1 and cis, for example, can be thought of as being established directly, or via the reverse of the other two processes:

$1 \rightleftarrows \text{cis} \qquad K_{1c}$

or

$1 \rightleftarrows \text{trans} \qquad 1/K_{t1}$

and

$\text{trans} \rightleftarrows \text{cis} \qquad 1/K_{ct}$

so we may write

$$K_{1c} = \frac{1}{K_{t1}K_{ct}} \tag{6.32}$$

Of course, this equation may be rearranged to give each one of the three equilibrium constants in terms of the other two.

When we now write rate constants for these reactions, the order of the subscripts can be used to indicate the direction of the reaction. For example, let

$$K_{1c} = \frac{k_{1c}}{k_{c1}} \tag{6.33}$$

where k_{1c} is the rate constant for conversion of 1-butene to *cis*-2-butene, and k_{c1} is for the reverse of this process. Since the equilibrium involves an equal number of moles of reactants and products, the orders of the forward and reverse reactions will be the same. Substitution of the six rate constants into Eq. (6.32) gives

$$\frac{k_{1c}}{k_{c1}} = \frac{1}{\dfrac{k_{t1}}{k_{1t}} \cdot \dfrac{k_{ct}}{k_{tc}}}$$

which can be arranged to a more symmetrical form,

$$k_{1c}k_{ct}k_{t1} = k_{c1}k_{tc}k_{1t} \tag{6.34}$$

That is, the product of the forward rate constants is equal to the product of the reverse rate constants. There is no theoretical requirement that the reaction orders for the three different chemical interchanges must be the same but, to anticipate a little, it does seem that all six rate processes are first order in hydrocarbon concentration, within the experimental range studied.

Without a catalyst, conversion among the three isomers is quite slow. The activation energy for the cis–trans isomerization, as was mentioned before, is about 65 kcal, while that for conversion to and from 1- to 2-butenes is even higher, so that temperatures over 400°C have been needed to give practical reaction rates.

Many different oxides and oxide systems have been tested as catalysts for these reactions. Two of the most interesting (9) have been aluminum sulfate and magnesium sulfate, both of these compounds being supported by SiO_2. That is, small SiO_2 particles were coated with the sulfates. The latter, being on the outside, acted as the catalysts. One important reason for using this technique is that the catalyst particles of each type have about the same size distribution, so that equal weights of catalyst have about the same total surface area—an important matter when comparing reaction rates.

The two compounds differ, in terms of adsorption and catalytic action, in that the aluminum ion is a strong Lewis acid, while the magnesium ion is

Table 6.4 Relative Rate Constants at 353°K for the Interconversion of Normal Butenes over Oxide Catalysts[a]

Oxide	k_{1c}	k_{1t}	k_{ct}	k_{c1}	k_{tc}	k_{t1}
Aluminum	1.00	1.02	1.29	0.22	0.54	0.09
Magnesium	1.00	0.81	0.26	0.22	0.11	0.07

[a] M. Misono and Y. Yoneda, *J. Phys. Chem.* **76,** 44 (1972).

a rather weak one. Although the catalysts were heated under vacuum before use, and were nominally anhydrous, in fact it would be expected that some water molecules would remain in the catalyst and would be held in such a way (the oxygens oriented toward the metal ions) that hydrogen ions would be available on the oxide surface. Because of the relative acidities, hydrogen ions would be more available on the aluminum sulfate than on the magnesium sulfate.

Experiments were carried out, starting with each of the three butenes. Rates of conversion over a given catalyst appeared to be proportional to the pressure of butene (first-order behavior) and *relative* first-order rate constants were determined, the data at 80°C (353°K) being given in Table 6.4. From the temperature dependence of the rate constants, activation energies for the total rate of conversion of each species (to both possible products) were determined, and are shown in Table 6.5. In general, aluminum sulfate was the better catalyst, a given amount producing two or three times as rapid reaction as magnesium sulfate. Aluminum sulfate was also more selective. When cis was the starting material, approximately six times as much trans as 1 was produced, and a corresponding selectivity of cis

Table 6.5 Activation Energies for the Interconversion of Normal Butenes over Oxide Catalysts[a]

	Activation energy (kcal) for conversion of		
Oxide	1-butene	*cis*-2-butene	*trans*-2-butene
Aluminum	7.2	7.5	8.0
Magnesium	9.5	10.0	11.8

[a] M. Misono and Y. Yoneda, *J. Phys. Chem.* **76,** 44 (1972).

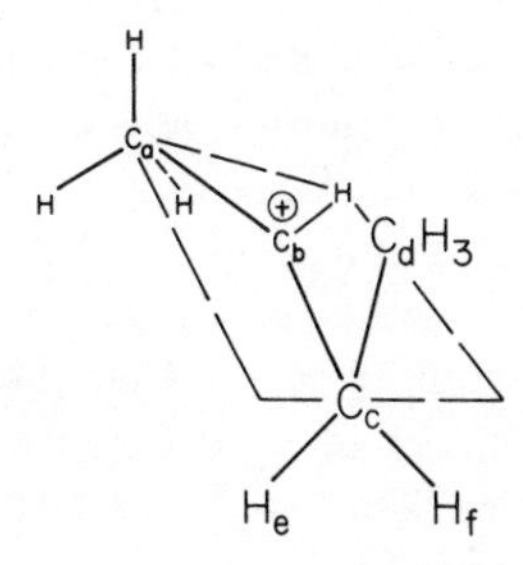

Fig. 6.3 Representation of the secondary carbonium ion proposed as the common intermediate in the isomerization of *n*-butenes (11).

over 1 was found when starting with trans (but not when starting with 1-butene). Magnesium sulfate, on the other hand, tended to convert any given butene into nearly equal amounts (within a factor of 1.6) of the other two, at the beginning of the reaction.

We might note that if the rates of reaction were governed strictly by equilibrium thermodynamic considerations, it would be expected that the ratios of products formed would simply be those at equilibrium, as calculated above. Starting with the cis isomer, the trans:1 ratio would be 0.70:0.04, or about 18:1. This ratio is not, of course, actually produced by either catalyst, so that it is clear that the rates are *not* determined only by thermodynamics. Equilibration of the reaction system, starting with cis, will actually occur by a somewhat roundabout process of first producing too much 1-butene and not enough trans, then conversion of the excess 1-butene to both cis and trans, and so on. Overall rate equations that show the dependence of each concentration on time have been worked out by Haag and Pines (10).

The symmetry of the total reaction system, and the fact that the rate constants and activation energies for all conversions on a given catalyst are fairly close to one another, led to the idea that the isomer conversions occur through a common intermediate, a carbonium ion formed by addition of a H^+ ion from the catalyst to the butene, the carbonium ion being adsorbed on the catalyst surface. In each case the ion would be a secondary one which can be formed from any of the butenes by addition of hydrogen at the right place in the molecule, followed by charge redistribution. Hightower and Hall (11) have shown a reasonable configuration for the ion, which is sketched in Fig. 6.3. The sketch is intended to show that carbons C_a, C_b and C_c, and the hydrogen attached to C_b, are approximately in the plane of the catalyst's surface. The two lower hydrogens of C_a and C_c are oriented downward slightly, fitting into molecular-sized irregularities in the catalyst surface (these need not be very big since H atoms are very small) while the upper hydrogen on C_a and the methyl group containing C_d project upward away from the catalyst surface. Loss of one of the lower H atoms attached to C_a will lead to the formation of 1-butene, loss of H_e will lead to *cis*-2-butene, while loss of H_f will lead to *trans*-2-butene. Consideration of the statistical factor only would indicate that the amount of 1-butene produced would be double that of each of the other isomers, but, since the hydrogens are not equivalent, there could well be energy effects too. In general, we would expect that loss of one of the secondary hydrogens, H_e and H_f, would be a lower energy process than loss of one of the primary hydrogens on C_a would be. One piece of direct evidence for this ionic intermediate is that when butenes are isomerized over a catalyst that contains deuterium instead of hydrogen ions, just one deuterium becomes associated with the hydrocarbon in the process.

Energy diagrams that fit the facts within an experimental error of about 0.5 kcal in activation energy are given in Fig. 6.4. The diagrams are qualitatively similar, but the higher acidity of the aluminum leads to somewhat lower activation energies for the formation of the ion from all the butenes, and a considerably lower energy of formation of the ion (the actual energies of the adsorbed ions relative to the gaseous butenes have not been determined—product ratios are indicative only of energy differences).

For the aluminum sulfate, the activation energies for formation of the 2-butenes from the ion are nearly the same, as they should be to give equal yields from 1-butene, as observed. The larger activation energy for the formation of 1-butene from the ion leads to a lower yield of it compared to either of the 2-butenes, even with a higher statistical factor. For magnesium sulfate, the slightly higher activation energy for formation of 1-butene from the ion combines with the higher statistical factor to give equal yields of all three butenes.

Finally, let us check the rate law given by this mechanism. If we start, for example, with pure 1-butene and consider the initial rates, the relevant chemical equations are

$$\begin{aligned} &1 + \text{cat} \underset{k_{-1}}{\overset{k_1}{\rightleftarrows}} \text{ion} \\ &\text{ion} \xrightarrow{k_2} \text{cis} + \text{cat} \\ &\text{ion} \xrightarrow{k_3} \text{trans} + \text{cat} \end{aligned} \qquad (6.35)$$

where "cat" stands for an unoccupied area on the catalyst, and "ion" for the adsorbed carbonium ion. We could also write, as we did for enzyme

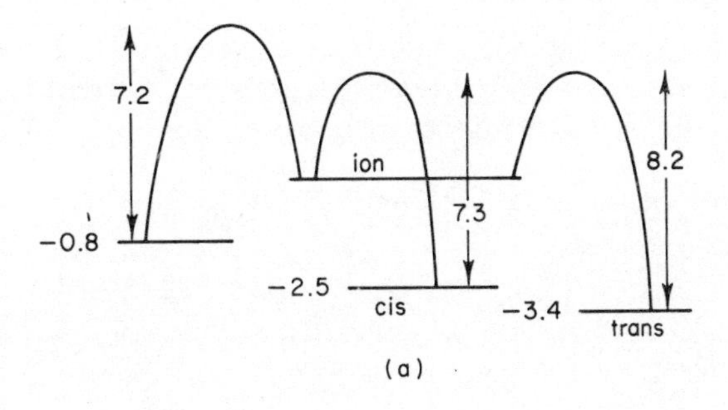

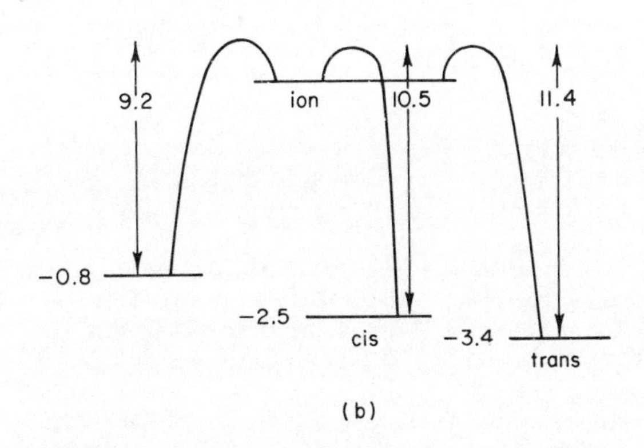

Fig. 6.4 A likely energy diagram for the conversion of normal butenes on metal sulfate catalysts ((a) aluminum sulfate, (b) magnesium sulfate) at 350°K. Energies of butenes are enthalpies of formation from Table 6.3. (Based on Ref. 9.)

catalysis,

$$[\text{cat}] + [\text{ion}] = [\text{cat}]_0$$

where, of course, we are talking about moles of catalyst sites and ions associated with a liter of gas, although not "in" the gas. Clearly, a steady-state calculation just like those of the Lindemann or Michaelis–Menten cases will work, the results of which are:

$$\text{rate}_{1c} = \frac{k_2[1][\text{cat}]_0}{[1] + (k_{-1} + k_2 + k_3)/k_1} \tag{6.36}$$

$$\text{rate}_{1t} = \frac{k_3[1][\text{cat}]_0}{[1] + (k_{-1} + k_2 + k_3)/k_1} \tag{6.37}$$

Since first-order behavior with respect to butenes was found, it appears that the experiments were done in the low-pressure region, so that the second term dominates in the denominator, and

$$\text{rate}_{1c} = \frac{k_1 k_2 [1][\text{cat}]_0}{k_{-1} + k_2 + k_3} \qquad (6.38)$$

$$\text{rate}_{1t} = \frac{k_1 k_3 [1][\text{cat}]_0}{k_{-1} + k_2 + k_3} \qquad (6.39)$$

Clearly, the overall rate of the process is governed by k_1, since the ratios $k_2/(k_{-1} + k_2 + k_3)$ and $k_3/(k_{-1} + k_2 + k_3)$ will be of the order of 2, and will not change much with temperature.

References

1. R. E. Carter and F. D. Richardson, *Trans. AIME* **203,** 336 (1955).
2. E. A. Gulbransen and K. Andrew, *J. Electrochem. Soc.* **98,** 241 (1951).
3. R. E. Carter and F. D. Richardson, *Trans. AIME* **200,** 1244 (1954).
4. C. Wagner, *Z. Physik. Chem.* **B21,** 25 (1933); see also the monograph by P. Kofstad in the Further Reading section.
5. W. P. Roe, H. R. Palmer and W. P. Opie, *Trans. Amer. Soc. Metals* **52,** 191 (1960).
6. I. Langmiur, *Trans. Faraday Soc.* **17,** 621 (1922).
7. K. G. P. Sulzmann, B. F. Myers, and E. R. Bartle, *J. Chem. Phys.* **42,** 3969 (1965).
8. I. I. Tretyakov, A. V. Sklyarov, and B. R. Shub, *Kinet. Katal.* **11,** 166, 479 (1970); in English translation as *Kinetics and Catalysis* **11,** 133, 397 (1970).
9. M. Misono and Y. Yoneda, *J. Phys. Chem.* **76,** 44 (1972).
10. W. O. Haag and H. Pines, *J. Amer. Chem. Soc.* **82,** 387 (1960).
11. J. W. Hightower and W. K. Hall, *J. Phys. Chem.* **71,** 1014 (1967).

Further Reading

A. Clark, "The Theory of Adsorption and Catalysis." Academic Press, New York, 1970.

N. B. Hannay, "Solid-State Chemistry." Prentice Hall, Englewood Cliffs, New Jersey, 1967.

J. A. Hedvall, "Solid State Chemistry." Elsevier, Amsterdam, 1966.

P. Kofstad, "High-Temperature Oxidation of Metals." Wiley, New York, 1966.

G. A. Somorjai, "Principles of Surface Chemistry." Prentice-Hall, Englewood Cliffs, New Jersey, 1972.

H. H. Uhlig, "Corrosion and Corrosion Control." Wiley, New York, 1963.

Problems

6.1 As pointed out in the text, derivation and use of the Wagner oxidation theory is awkward when the traditional units are used. Redefine a "rational rate constant" in the SI units as the number of moles of oxygen added to the oxide layer per square meter of surface per second, for a thickness of 1 m. Then rederive Eq. (6.14) using SI units in all of the constituent relationships. Show, by working out conversion factors for the various quantities from the traditional to SI units, that your new equation gives correct numerical answers.

6.2 For the metal chromium, Gulbransen and Andrew (*J. Electrochem. Soc.* **104,** 334 (1957)) found that the parabolic rate constant for oxidation is 8.2×10^{-15} $(g/cm^2)^2/sec$ at 700°C. and 0.1 atm oxygen pressure. They also found that the rate constant depended on temperature according to an Arrhenius rate law with an activation energy of 37,500 cal.

(a) If it is assumed that the oxide formed is Cr_2O_3, and that chromium metal has a density of 7.1 g/cm^3 with a negligible change with temperature, convert the given rate constant into one that would give the thickness of chromium metal converted to oxide per unit time. As a parabolic constant, the units of your quantity could be cm^2 sec^{-1}.

(b) Suppose that in a certain high-temperature application of chromium at the same oxygen pressure, no more than 1 mm thickness of chromium should be oxidized over a three-year period. What would be the maximum temperature at which the chromium could be used?

6.3 The rate of dissolution of solids may be limited by diffusion if there is no stirring and if the geometry is such that convection currents are not set up due to density gradients. As an example, suppose that a smooth layer of small crystals of a salt in the bottom of a beaker is covered with several cm of water. The solution adjacent to the crystals would soon become saturated, while the concentrations above the salt layer would increase relatively slowly with time, approximately in accordance with Eq. (6.20). Approximations enter because the location of the salt–solution interface moves downward as the salt dissolves, the volume of an element of water changes as salt dissolves in it, and D depends slightly on composition. (Can you think of any other sources of approximations?) Equation (6.20) will be approached for a slightly soluble salt. In this physical situation, convection will be minimized because the most dense solutions will be at the bottom.

For NaCl, the diffusion coefficient is within a few percent of 1.5×10^{-5} cm^2 sec^{-1} over the concentration range from 0 to saturation (5.8 M) at room temperature. Calculate the concentrations of NaCl in the 3 cm solution above the salt layer at several different times, comparing your results with Fig. 6.1.

6.4 An interesting discussion of the decomposition of $CaCO_3$ to CaO and CO_2 is given by I. B. Cutler (in "Kinetics of high-temperature processes," W. D. Kingery, ed., p. 294, Wiley, New York, 1959). Since $CaCO_3$ and CaO have different crystal structures, he assumes that CaO crystals form near, but separate from, a reacting $CaCO_3$ surface. As CaO molecules form, they momentarily remain on the surface, covering a fraction θ and preventing further reaction at that point, but they soon diffuse to a growing CaO crystal, and no longer interfere with the reaction. The CaO molecules on the $CaCO_3$ surface are denoted CaO*, since they would be expected to be at higher energy than if they were part of a CaO crystal. The reaction sequence, then, is

$$CaCO_3 \underset{k_{-1}}{\overset{k_1}{\rightleftarrows}} CaO^* + CO_2(g)$$

$$CaO^* \underset{k_{-2}}{\overset{k_2}{\rightleftarrows}} CaO$$

The rate constants apply, as Langmiur used them, to the situation where all of the surface is available for the reaction to occur.

At a steady state,

$$\begin{aligned} \text{rate} &= -\text{rate of change of weight of sample per unit time} \\ &= k_0[k_1(1-\theta) - P_{CO_2}k_{-1}\theta] \end{aligned}$$

where k_0 is a constant that takes care of the amount of substance present. Also,

$$\frac{d\theta}{dt} = k_1(1-\theta) - P_{CO_2}k_{-1}\theta - k_2\theta + k_{-2}(1-\theta)$$

(a) Using the steady-state approximation, show that

$$\text{rate} = k_0 \frac{k_1k_2 - k_{-1}k_{-2}P_{CO2}}{k_1 + k_2 + k_{-2} + k_{-1}P_{CO_2}}$$

(b) Obtain a limiting rate constant at low P_{CO_2} pressures.
(c) Show that if the CO_2 pressure is set at the equilibrium value,

thermodynamic requirements on the rate constants cause the rate given in part (a) to go to zero, as it should.

(d) At 900°C, the equilibrium CO_2 pressure is approximately 1 atm. Draw, qualitatively, a sketch of how the rate of reaction would depend on CO_2 pressures between 0 and 1 atm at 900°C.

6.5 As part of the results of his studies of the interaction of gases with solids, Langmiur developed an equation for the fraction of surface covered by adsorbed gas that is frequently called the "Langmuir isotherm." For adsorption of a substance on a surface to form a layer of adsorbed molecules no more than 1 molecule thick, only a single equation is necessary

$$A + S \rightleftarrows AS$$

where A is a gas molecule, S a surface site, and AS an adsorbed molecule. Using Langmiur's approach as described in the text, show that

$$\theta = \frac{KP}{1 + KP}$$

where θ is the fraction of surface covered, P the pressure, and K the adsorption equilibrium constant, which is equal to $\mu/\nu P$, where μ is the rate at which molecules strike 1 cm² of surface, and ν is the rate at which they leave 1 cm² of completely covered surface.

6.6 In one of his experiments, Langmuir measured the initial rate of the reaction

$$2\ CO + O_2 \rightarrow 2\ CO_2$$

catalyzed by platinum at 569°K and pressures of 3.0×10^{-6} atm CO and 1.4×10^{-6} atm O_2, obtaining a value of 1.6 mm³ min cm². His rate is measured in terms of the number of cubic millimeters of oxygen, measured at 298°K and 1 atm, that react with CO per minute on 1 cm² of platinum.

(a) Convert this rate into the quantity we have called W, the number of moles of O_2 reacted per second per cm² of surface.

(b) Calculate the quantities μ_1 and μ_2, the numbers of moles of CO and O_2 striking 1 cm² of surface per second.

(c) From Eq. (6.27), calculate the value of ν_1 at this temperature.

(d) Compare the terms in the denominator of Eq. (6.27). Do these conditions correspond to a limiting or an intermediate case? Can you write a rate law that would apply (approximately) under these conditions?

6.7 Write a computer program to calculate W, the rate of the platinum-catalyzed reaction between CO and O_2, as given by Eq. (6.27). The necessary values of μ_1 and μ_2 can be set up as functions of pressure and temperature, while ν_1 can be calculated from the data of Problem 6.6 and the activation energy found for the loss of CO from a platinum surface of 34 kcal, given in the text. Use your program to calculate several reaction profiles. For example, choose fixed CO and O_2 pressures and vary the temperature, or vary the CO/O_2 ratio at fixed temperature and total pressure, etc.

6.8 Find out what CO and O_2 concentrations are present in the exhaust gases of a typical car, and the volume of gases produced per second. If one used a catalytic reactor (with platinum as the catalyst) of volume 0.5 ft^3, suggest some combinations of catalyst area and reaction temperature that could remove 90% of the CO from the exhaust gases, based on Equation 6.27.

6.9 The decomposition of N_2O on platinum has been studied under various conditions of temperature and pressure over a period of many years. Redmond (*J. Phys. Chem.* **67**, 788 (1963)) made a study at relatively low N_2O pressures, and in the presence of added O_2. He obtained an empirical rate law

$$\text{rate} = \frac{aP_{N_2O}}{1 + bP_{O_2}{}^{1/2}}$$

Apparently, oxygen (which might be added at the beginning of the experiment, or might simply be a reaction product) inhibits the reaction, presumably by occupying reaction sites on the surface. Show that, using Langmuir's approach, the sequence of reactions

$$N_2O(g) + S \rightleftarrows N_2O(ads)$$

$$N_2O(ads) \rightarrow N_2(g) + O(ads)$$

$$2\,O(ads) \rightleftarrows O_2(g) + 2\,S$$

where S stands for an adsorption site, will lead to the above rate law, given some reasonable assumptions. These are (a) N_2O is weakly adsorbed on platinum, so that the fraction of surface covered by it is very small compared to the fractions uncovered or covered by O. If this fraction is θ_1, then θ_1 and terms containing it may be neglected when they are combined with other terms by addition or subtraction. (b) When two adjacent lattice spaces are involved in a reaction, the probability that both will be occupied by reacting molecules or that both will be unoccupied is the square of the probability that one will be. For example, if θ_2 is the fraction of surface covered by O atoms, and ν_2 the rate of loss of O atoms from unit surface that is completely covered, then the rate of loss from a partially covered surface is $\nu_2\theta^2$.

6.10 In a study of the isomerization of cyclopropane to propylene over a silica-alumina catalyst, Hightower and Hall (*J. Phys. Chem.* **72**, 4555 (1968)) carried out a series of experiments to determine empirically the order of the reaction. In a flow reactor at 150°C, they obtained the following initial rates of reaction, on a relative basis:

Initial P (Torr)	Initial rate, relative
6.8	1.00
18.6	1.04
24.1	1.20
59.7	2.09
65.8	2.11
80.7	2.26

From these data, obtain the value of n that would apply to a rate equation

$$\text{rate} = kP^n$$

Compare your result with the author's value of 0.5 for n. Suggestion: Can you think of a convenient way of plotting the data so as to get a straight line with slope n?

7 Experimental Methods

What is the best way to measure the rate of a chemical reaction? That depends, of course, on the reaction in which we are interested; but most methods can be classified as either static or flow methods, with a second classification into slow and fast methods that is not fundamental, but made only for convenience because of the need for special instruments to follow rapid changes. Accordingly, we will discuss experimental methods under the above headings.

7.1 STATIC METHODS FOR SLOW REACTIONS

7.1.1 General Considerations

In an ideal static reactor for kinetic studies we would like to bring a reaction mixture (or single reactive substance) instantaneously from a completely unreacted state to a known condition of concentration, temperature, and pressure, and then to make a series of analyses of the mixture at known intervals of time, while maintaining the temperature and pressure exactly constant. Often these conditions can be well approximated, while sometimes (as when we know that the total pressure will have little effect on the rate of the reaction) we may let one or more conditions vary, or even use the variation as a method of following the reaction.

7.1.2 Reactions in Solution

Let us first look at the method used to obtain most of the kinetic data on organic reactions, such as those correlated by the Hammett equation. A kinetic study of the base hydrolysis of ethyl benzoate by NaOH, for example, would probably be carried out in a variety of mixed solvents such as ethanol–water or acetone–water, in order to test the effect of dielectric constant and other solvent effects. The experimental range would be limited by the volatility of the solvents, and the solubility of the reactants in them. Since pressure has been found to have little effect on these reactions, atmospheric pressure has been almost universally used. But what is the best way of getting the reaction started?

One approach is to prepare a reaction mixture at a low temperature, at which the rate will be very slow, then bring it up to the reaction temperature by placing it in a large water bath which is at the desired temperature. For liquids, because of the finite time necessary to transfer heat from the bath to the reaction mixture (usually through a glass flask) this method is not usually as good as bringing solutions containing the right amount of each reagent separately to the reaction temperature, then mixing the solutions at the appropriate time. That is, the mixing process in this instance is typically faster than the heating process, and it can be short enough (a few seconds) to be instantaneous compared to a total reaction time of thousands of seconds.

The reverse process of cooling samples taken during the reaction period can be done rapidly, for example, by adding the sample to ice water, and this method of "quenching" samples to keep them from reacting further during analysis is often used. It is quite effective in the above example, where the decrease in concentration of NaOH during the hydrolysis of ethyl benzoate may be followed by titrating a small, quenched sample of reaction mixture with HCl. It would seem to be even better if the reaction mixture could be analyzed in place by absorption spectroscopy, electrical conductivity, or some such method, and this is often done. It is not always the best approach, though, for the problem of maintaining a uniform temperature in a heated infrared cell, with its requirement for open light paths, may introduce more error than that of quenching a sample from a well-regulated water bath.

7.1.3 Reactions of Gases

Static methods also work nicely for slow gaseous reactions, the study of Schneider and Rabinovitch (1) discussed in Chapter 4 being a good example. Since the concentrations of gases at normal pressures are much less than those of liquids, the temperature of a gas sample can be raised with the addition of relatively little heat, so that it is often practical to start the reaction by adding a cold gas sample to a heated, evacuated bulb, the sample coming to the reaction temperature in a few seconds.

If the heating time still seems too long, the fact that gases have low viscosities enables one to introduce them into a reactor through a long, small-diameter but thick-walled tube which is at the reaction temperature, the gas being heated rapidly by contact with the tube wall. Actually it is just as well to have the gas enter the reactor at a temperature a little less than the desired temperature because when the gas comes to rest in the reactor its translational energy along the addition tube (which may be an appreciable fraction of the average speed of the molecules) is converted into random thermal energy, with a corresponding rise in temperature.

7.1.4 Heterogeneous Effects

More often than with solution reactions, the rates of gaseous reactions may be supplemented by heterogeneous reactions catalyzed by the container walls. Because the fraction of wall collisions increases as the pressure is lowered, heterogeneous reactions are typically most important at low pressures.

One common method of testing for heterogeneous reaction is to study the process in reaction vessels of various surface/volume ratios. A rate constant that is independent of vessel shape is probably homogeneous, while if a small variation of rate constant with vessel geometry is observed, the rate constant for the homogeneous reaction may be found by extrapolating mathematically to a surface/volume ratio of zero (that is, in effect, to a reaction vessel of infinitely large size). Experimentally, large reaction vessels over 200 ℓ have sometimes been used in low-pressure gas reactions, in order to minimize the surface/volume ratio.

Another approach is to change the material of the reactor wall. Perhaps some substances will have a much different catalytic effect than others. In his studies on the hydrogen–oxygen reaction, referred to at the end of Chapter 4, Hinshelwood used glass reactors, and also reactors coated with KCl and B_2O_3, finding large differences in the rates of the heterogeneous reactions. Cleaning his glass surfaces in different ways led to different catalytic effects, and even using a given reactor for a while changed its catalytic behavior. Because of all these surface effects, Hinshelwood was unable to eliminate surface effects completely at low pressures, and, as indicated in Chapter 4, his studies could not be made to yield good values of the elementary reactions in the hydrogen–oxygen reaction.

Finally, there have been examples where heterogeneous reactions have been shown to occur even when variations of the surface/volume ratio have not changed the apparent rate constant. In the first static bulb experiments on methane decomposition, rate constants higher than the true homogeneous ones were reported (2, 3). It was not until some time later that the reason for this was found to be the formation of small particles of carbon suspended in the gas; these particles catalyzed the reaction but were not directly related to the physical surface of the reaction vessel.

7.2 STATIC METHODS FOR FAST REACTIONS

7.2.1 Relaxation Methods

Over the past seventy years a whole family of methods for studying fast reactions in solution has been developed. The leader in the field has been Manfred Eigen of the University of Göttingen, Germany, who received the Nobel prize for this work in 1967. All of these techniques are relaxation methods (Section 5.5) in which small perturbations are produced in a solution at equilibrium, the approach to the new equilibrium being followed by fast-response instruments.

Probably the most common methods of following fast reactions are electrical conductivity and absorption spectroscopy. In the former, the reaction vessel, which of course must contain two electrodes, is made one arm of a Wheatstone bridge. Since there will hardly be time to balance the bridge during an experiment, the bridge is balanced beforehand, then the off-balance current or voltage measured and the change in resistance of the reaction cell calculated.

The small, rapid changes in off-balance voltage are easily measured by a sensitive, high-speed oscilloscope. For those who are used to seeing oscilloscopes present only continuous wave forms by repetitive sweeps of the electron beam, we might mention that most modern oscilloscopes can be set to make a single sweep when electrically triggered. Advances in oscilloscope performance in recent years have been dramatic; instruments that can measure signals of a millivolt or less, which appear and disappear in a few nanoseconds, are commercially available, the price of a good one being from $1000 to $4000.

In the absorption spectroscopy method, the shortness of time again leads to a variation from the usual approach. Rather than try to scan a spectrum in a microsecond, one sets the spectrometer to an absorption band of a reactant or product and monitors the change of intensity with time. Sometimes a double beam method is used, to balance out changes in lamp intensity and other instrumental variations that would produce spurious changes. At the present time, photomultiplier tubes of high sensitivity, stable power supplies for them, and high-performance oscilloscopes for displaying the data make this a relatively easy and effective method. Finally, it should be mentioned that permanent records of these oscilloscope traces are usually made photographically, Polaroid film being used so that the results of each experiment may be examined to help choose the right settings for the following one.

Let us now look at some of the techniques used to perturb solution samples to carry out the relaxation method. In one, the temperature of a sample is raised rapidly by discharging a capacitor through it. The time necessary to discharge the capacitor depends on its shape and the overall electrical characteristics of the circuit, and in the earlier designs was a few microseconds. Recently, by using a coaxial cable as capacitor, coupled with a very small sample of liquid, Hoffman (4) was able to raise the temperature of a solution 3° C in a few nanoseconds. The passage of the current causes some electrolysis of the sample; but, because of the short time, diffusion of the ions from the center of the cell does not occur; so as long as subsequent measurements are made near the center, the electrical method of heating produces minimal side effects. As an example of the application of this method, Eigen and co-workers (5) used it to measure the rate constants for the removal of a proton from a large number of analytical indicator reagents by OH^-. Concentrations of OH^- were low (that is, in the regions where the indicators changed color) and solutions were made electrically conducting by the presence of a small amount (0.1 M) of KNO_3.

The temperature may also be raised by exposing the specimen to a sudden pulse of radiation. Microwave radiation is especially useful in heating organic specimens that have too high electrical resistance to be heated by the passage of an electric current. Typically, microwaves are absorbed as rotational excitation, and rotational energy is quickly converted to other forms of thermal energy. Visible radiation may also be used, although many substances do not absorb these frequencies very well, and there is also the possibility that quanta of this size may produce chemical reactions directly, in which case the entire process studied will be different than what would be the case if the energy went into heat. One way to encourage heat production is to suspend small, inert, opaque particles in the liquid, which itself need not absorb the radiation.

The use of light pulses has increased since the development of "Q-switched" lasers (6). These lasers are triggered by motion of a mirror or by use of a Kerr cell so that all of the energy stored up is released in a burst that can have a time duration as short as a nanosecond. At present this is probably the fastest way of producing a sudden change in a chemical system, although some of the other methods are not far behind.

Some reactions of ions can be studied by subjecting a solution to sudden changes in electric field, since the dissociation constant of electrolytes depends slightly on the applied field (7). Since strong fields, of the order of 100 kV/cm, are required, the method works best with solutions of low conductivity. Sometimes a single voltage pulse is applied, while it is also possible to apply a square-wave potential and measure a series of perturbations. The "E-jump" method was one of the first applied by Eigen and co-workers, being used to measure the rate constants for the combination of hydrogen ions with hydroxyl and acetate ions (5).

For the study of reactions involving volume changes, the method of suddenly changing the pressure has proved valuable. Usually this has been accomplished by transmitting a high-frequency sound wave through the solution. The frequency of the sound is varied, and when the frequency approaches the relaxation time τ for a chemical process, the sound is absorbed by the solution—that is, the pressure changes that make up the sound wave are reduced as the solution converts back and forth, to and from the more- or less-voluminous state in response to the stress of the sound wave. This is an easy method to use since fast-response instruments are not needed: only transducers to produce the sound and detectors to measure it. On the other hand, there is no clear way of telling what chemical change produced the sound absorption and often the technique will have to be supplemented by other methods to characterize the chemical process being studied.

The pressure-variation method has been specially useful in the study of hydrated and other complex ions in solution. In the case of aqueous solutions of copper sulfate, for example, it has been found (8, 9) that solutions consist not only of individual hydrated copper and sulfate ions that interact with each other electrostatically over relatively large distances (the Debye–Hückel picture of a dilute electrolyte) but also of two other species. One of these consists of a pair of ions that have come together and associated, each maintaining its complete hydration sphere (an outer-sphere complex), the other species being an associated pair of ions that has lost one or more molecules of water so that the ions can approach one another more closely to form an inner-sphere complex. The equilibrium between these two types of complex is pressure dependent, and by varying the pressure in solution ultrasonically the rate constant for loss of a water molecule from an outer-sphere complex has been found to be 10^9 sec^{-1}.

Larger pressure jumps can be brought about by single-step methods to study reactions for which the dependence on pressure is fairly small. In one technique a sample is pressurized in a container, one side of which is a rupture disk. When the disk breaks, the pressure drops quickly to atmospheric, since the volume change is small and the speed of sound in liquids high. For example, in a typical liquid such as ethanol the speed of sound is about 10^6 cm/sec, so that a pressure change would propagate through a solution in 10^{-6}–10^{-5} sec. A liquid may also be subjected to a sudden increase of pressure in about the same time, if the sample is placed in the end of a tube in which a gaseous shock wave is produced (Sec. 7.4.2).

7.2.2 Nuclear Magnetic Resonance

Kinetic data on reactions in solution can be obtained by nuclear magnetic resonance (NMR) methods. The basis of these methods is the fact that when two compounds with different NMR chemical shifts convert rapidly from one to the other, their two NMR peaks merge into one (10).

The type of behavior found is shown in Fig. 7.1. In this figure, we have assumed for simplicity that there is no splitting of the NMR lines, that the equilibrium constant for the reaction $A \rightleftarrows B$ has the value 1 at all temperatures, and that therefore the rate constants and relaxation times for the forward and reverse reactions remain equal to one another at all temperatures. We do assume, however, that the rate constants (and relaxation times) may vary with experimental conditions—to be specific, we may assume that the rate constants increase with increasing temperature. Such restrictions, of course, would not be expected to hold in real experimental studies, so actual NMR data will look much more complicated than those Fig. 7.1. However, methods of handling the complications have been

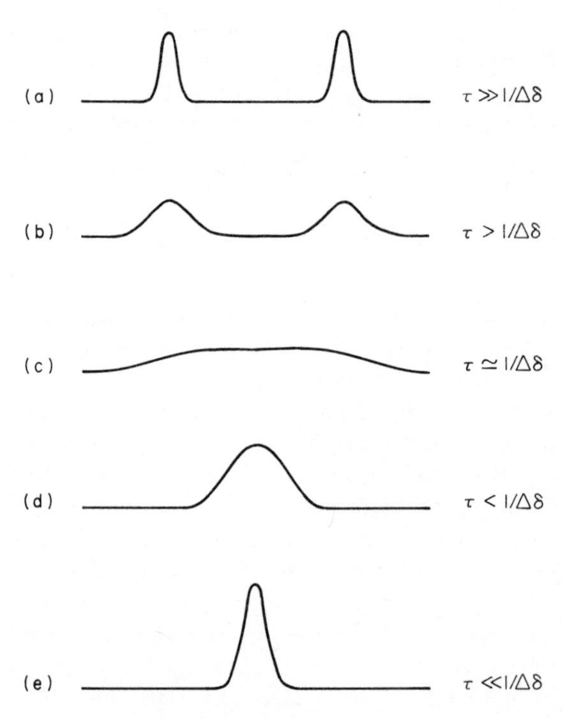

Fig. 7.1 Combination of two NMR peaks as the rate constant for chemical conversion increases.

developed to the point where a wide variety of NMR patterns can be analyzed kinetically.

In Fig. 7.1a are shown the two peaks due to the two species, at low temperature where the difference in their chemical shifts ($\Delta\delta$) is much larger than the (first-order) rate constant for their interconversion. In terms of the relaxation time τ for interconversion, the condition in (a) is that $\tau \gg 1/\Delta\delta$. In typical NMR apparatus, the quantity $\Delta\delta$ would be expected to be of the order of 10–100 Hz, the actual value depending, of course, on the reacting substances. As the temperature rises and τ decreases, the lines broaden, as in Fig. 7.1b, and when $\tau \simeq 1/\Delta\delta$, they merge into a single broad peak (c). Finally, as in (d) and (e), the peak sharpens until, when $\tau \ll 1/\Delta\delta$, a peak resembling that for a single substance is obtained.

A typical example of the method is described by Pinnavaia and Lott (11). They studied the NMR spectra of (among other compounds) a zirconium complex, (π-cyclopentadiene)-Zr-(acetonylacetone)$_2$Cl in benzene. The complex, which exists in two stereoisomers, showed a pair of clearly defined peaks, one from each isomer, separated by about 3 Hz. These remained sharp from room temperature to 65° C, then went through a transition very similar to that of Fig. 7.1, merging at 87° C and becoming a sharp single peak above 100° C. Calculated relaxation times varied from 0.26 sec at 76° C to 0.026 sec at 100° C. The small $\Delta\delta$ in this case was due to the small difference in environment of the proton caused by the chemical change.

The NMR method is in some ways a relaxation one, in that the systems studied are usually close to equilibrium, those molecules which are excited by the absorption of radiation losing their excess energy rather quickly in collisions. As with the relaxation methods, rate constants of second-order reactions with numerical values substantially above $\Delta\delta$ may be studied when the concentrations of reactants are substantially below 1 M (Problem 7.7).

7.2.3 Stopped Flow

One widely used static method for liquids was developed from a flow method, and is called the "stopped-flow" method. The idea is illustrated in Fig. 7.2. In this technique, two reacting solutions are placed in separate hypodermic syringes, the flows from the two syringes being brought together in a small mixing chamber and then allowed to flow out down a single tube in which there is an observation port. After the flow has continued for a short while it is stopped by an obstruction downstream from the viewing port, and the sample that happens to be in the viewing region at this time is analyzed over a period of time, typically by absorption spectroscopy.

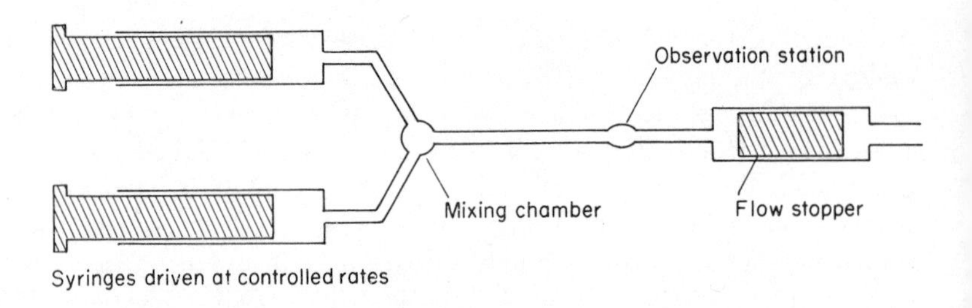

Fig. 7.2 Schematic of stopped-flow apparatus.

This method is not applicable to the fastest solution reactions since the time to mix the solutions and move them a short distance from the mixing chamber is seldom less than a millisecond. However, it has been found that there are many reactions (including a large number of enzyme and other reactions of biological importance) that can be studied by this technique, so that it has been found worthwhile to build highly automated stopped-flow reactors (12), including some that are available commercially.[1]

7.2.4 Flash Photolysis

For studying fast gas reactions a method that has a superficial resemblance to the family of relaxation methods is flash photolysis. Developed by Drs. Ronald Norrish and George Porter in England at the same time that Eigen was doing his pioneering work on liquids (and leading to Nobel prizes for Norrish and Porter along with Eigen), the method consists of subjecting a gas sample to a short, intense pulse of light that starts chemical reactions, then analyzing the gas sample spectroscopically after a short, known, interval of time. A typical experimental arrangement is shown in Fig. 7.3.

The principal difference between this and the relaxation methods is that the amount of energy put into the gas is relatively large, so that the temperature of the gas is raised by 1000° or more in a few microseconds or, if the energy is absorbed into a particular degree of freedom in the molecules, highly excited species or dissociated species are produced. In one of the first applications of the method (13) chlorine–oxygen mixtures were flashed, the radiation dissociating a substantial fraction of the Cl_2 to Cl

[1] American Instrument Co., 8030 Georgia Ave., Silver Spring, Maryland 20910; Durrum Instrument Corp., 925 East Morrow Drive, Palo Alto, California 94303.

atoms, other reactions following to produce the sequence:

$$Cl_2 \rightarrow 2\ Cl$$

$$2\ Cl + O_2 \rightarrow 2\ ClO$$

$$\rightarrow Cl_2 + O_2$$

$$2\ ClO \rightarrow Cl_2 + O_2$$

The ClO was produced very quickly, and its presence and rate of disappearance were followed by taking a series of absorption spectra (one per experiment) at times up to 10 millisec after the initial flash. Here, the translational and rotational temperature of the molecules stayed close to ambient, the energy being specifically absorbed in the Cl—Cl bond. Later, using a faster flash, Nicholas and Norrish (14) were able to measure the rate constant for the addition of Cl atom to O_2, finding it (not surprisingly) to be third order, with a rate constant of 6×10^{14} mole^{-2} cc^2 sec^{-1}, where N_2 is the principal third body.

In other experiments, such as those of Erhard and Norrish (15) on the oxidation of hydrocarbons, the energy absorbed produced both direct dissociation of molecules and heating of the gas, so that further reactions of the radicals produced occurred in a high temperature environment. If this division of energy occurs, it is not always easy to obtain the temperature of the gas after the flash, or to calculate how it changes subsequently. Accordingly, the best understood results obtained by this method are those in which the light energy produced atoms or radicals, and their subsequent reactions were followed at, essentially, room temperature. Recently, Porter and Topp (16) have used a pulsed laser to produce photolysis flashes of 25 nanosec duration, and study the absorption spectrum of species with half-lives near 100 nanosec.

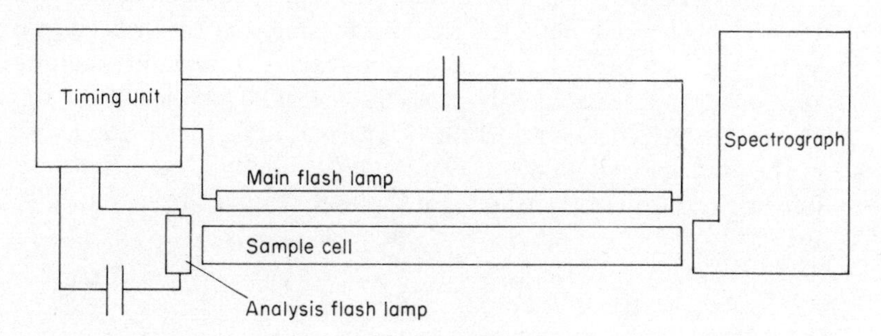

Fig. 7.3 Schematic of flash photolysis apparatus.

7.2.5 Steady Photolysis

Probably it would not be fair to omit steady-state photolysis as a method of studying fast reactions. In this technique, a gas sample is irradiated for a long time (typically hours) with radiation that produces excited species. Frequently the excited species is allowed to react with two kinds of molecules, the rate constant with one of them being known, and the rate constant with the other being calculable from the ratio of product molecules. In this way fast reactions may be studied in a relatively simple experiment.

7.2.6 Shock Tube

The shock tube, another technique for studying gas reactions, can be operated as either a static or flow reactor, and we will discuss it under the latter heading.

7.3 FLOW METHODS FOR SLOW REACTIONS

7.3.1 Tubular Reactors

It would seem a simple matter to measure the rate constant for a reaction by passing a solution or gas mixture through a tubular reactor that is at a certain temperature. Analyses of the reaction mixture before entering and after leaving the tube would give the required concentration changes, while the residence time in the reactor would simply be equal to the volume of the reactor divided by the flow rate. In fact, while tubular flow reactors are fairly commonly used for kinetic studies, the complexities of fluid flow make the calculation of rate constants more difficult than might be expected.

Most commonly, the flow of fluid through a tubular reactor is *laminar*. Flow is most rapid along the axis of the tube, the flow velocity being zero at the tube wall. Laminar flow is usual when two conditions are fulfilled: (a) the mean free path of the molecule is a small fraction of the tube radius (typically, pressures above 1 mm for gases); and (b) the Reynolds number of the fluid is less than 3000. The Reynolds number, symbolized by Re, is a dimensionless quantity which, for the flow of fluid through a tube, is given by

$$\mathrm{Re} = \frac{dv\rho}{\eta}$$

where d is the diameter of the tube (cm), v the average flow velocity

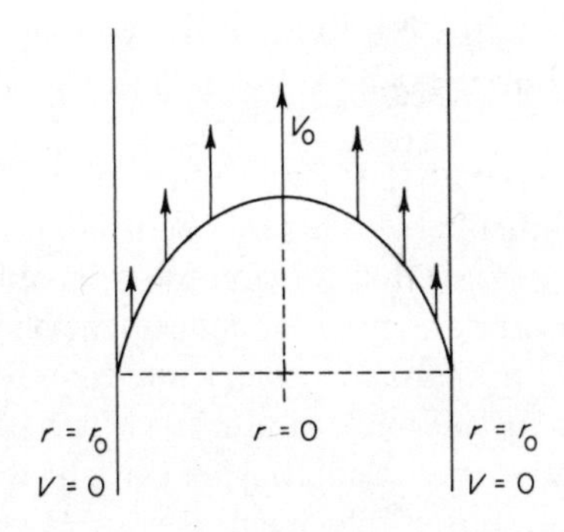

Fig. 7.4 Parabolic velocity distribution in laminar flow.

(cm/sec) obtained by dividing the flow rate by the cross-sectional area, ρ the density of the fluid (g/cc), and η the viscosity of the fluid (poise). As an example, suppose a tubular reactor is made from tubing 0.5 cm in diameter and 60 cm long, with an average flow velocity of 12 cm/sec so that the average residence time is 5 sec, and that the fluid is water with a viscosity of 1 centipoise (0.01 poise). Then

$$\mathrm{Re} = \frac{0.5 \times 12 \times 1}{0.01} = 600$$

For gases, both density and viscosity will be lower, typical numerical values leading to laminar flow in many typical experimental situations.

Laminar flow has a characteristic parabolic velocity distribution as a function of distance from the center of the tube (Fig. 7.4). With reference to the figure, the flow velocity at some radius r is given by

$$V_0 - V = kr^2$$

To match the condition that the velocity should be 0 at the tube wall,

$$V_0 = kr_0^2$$

so that

$$k = \frac{V_0}{r_0^2}$$

and

$$V = V_0\left(1 - \frac{r^2}{r_0^2}\right)$$

This equation may be integrated over the tube radius to show that the apparent or mass flow velocity (calculated by dividing the volume of flow per unit time by the cross-sectional area of the tube) is half the velocity at the center.

If the molecules of fluid always stayed at the same radius as they passed through the tube, it would not be difficult to calculate the residence times of molecules at different radii, and this time distribution would enable us to calculate rate constants from the reactant or product concentrations in the fluid leaving the reactor. Actually, a realistic calculation requires that radial diffusion of molecules be taken into account, and while this is possible, as Poirier and Carr (17) have shown, the calculation is complicated and time-consuming, requiring use of a computer. The result is worth the effort, though, if one is interested in reactions for which the time scale is in the range of 0.1 to 100 sec—times that are long in terms of most relaxation or other "fast" methods, but shorter than can conveniently be studied by the older static methods. In particular, many industrial processes are carried out in reactors of this type, so that the industrial research chemist will often find that a tubular flow reactor will not only give him results in the range of interest, but also under the conditions liable to occur in the plant that (he hopes) will be built as the result of his efforts. For example, he may find that a certain by-product tends to form on the tube walls because of the relatively long reaction time in that section, maybe a solid that would tend to plug up a commercial reactor after a month's operation. He might not have seen this product in a static reactor, because of the absence of a residence time distribution.

Tubular flow reactors are commonly used to study catalytic reactions. It is easy to fill a vertical tube with catalyst pellets and carry out the reaction. Some of the differences that should be thought of when a catalytic study is being made are: (a) The cross-sectional area of the tube is reduced, so the residence time for a given total flow rate is less with catalyst in place than without it. (b) At a given flow velocity, the tendency toward laminar flow will be greater since the diameter of the spaces through which the gas is flowing will be smaller. At a given flow rate, the higher velocity may lead to turbulent flow. (c) The catalyst particles will tend to even out the flow velocity across the tube since in effect there will be many small tubes instead of one large one. The radial mixing due to random particle orientation will tend to keep radial concentration gradients small. For these reasons, flow in a packed reactor will approach "plug flow," where molecules entering at all radii will experience similar dwell times in the reactor. (d) If the catalyst is porous, a variation in residence time may arise because some molecules will happen to stay absorbed longer than others. This happens to a lesser degree by absorption on the surface of a solid catalyst, or in the liquid film of a supported liquid catalyst. Anyone who has operated a gas chromatograph has noticed how a sample, injected at one time, becomes distributed along the column by this type of process.

At low gas pressures, where the first requirement for laminar flow is not fulfilled, molecules will diffuse freely in a radial direction in the tube, while maintaining an average flow velocity along the tube. The distribution of residence times in the reactor, in this case, will depend on the length/diameter ratio of the tube, and on the ratio of flow to molecular velocities, and can be calculated from the kinetic theory. In this situation the molecules will experience a relatively large number of collisions with the wall, so heterogeneous reactions may be important.

If the Reynolds number exceeds 3000, *turbulent flow* may be expected. Here the fluid swirls as it flows along the tube, leading to much more radial mixing than for laminar flow, so that the velocity profile across the tube is flatter than occurs for laminar flow. Typically, the apparent mass flow velocity is 80% that of the flow velocity along the center—almost close enough to allow one to neglect the radial variations. Actually, the transition from laminar to turbulent flow is not a sharp one, and the value of Re at which it occurs depends on the material of the tube wall, so that it is generally undesirable to carry out experiments in the range $1000 < \mathrm{Re} < 10{,}000$, in which the type of flow may not be clearly understood.

7.3.2 Well-stirred Reactor

Another type of flow reactor that is an analogue of an industrial type is the "well-stirred reactor." Here, one has a container into which the reactant flows at a steady rate, while an equal stream of fluid is withdrawn from another point in the reactor (Fig. 7.5). It is assumed that the stirring is efficient, so that the stream being withdrawn is typical of the entire contents of the reactor.

The mathematics of this system is simpler than for a tubular reactor. All we need do is write a material balance for a reactant:

rate of inflow − rate of reaction × volume of reactor
= rate of outflow

Let us do an example using a familiar hydrolysis reaction. Suppose that we operate a well-stirred reactor under the following conditions:

Inflow: 1 ml/sec of 0.1 *M* NaOH
1 ml/sec of 0.1 *M* Ethyl acetate
Outflow: 2 ml/sec of solution containing 0.02 *M* of both NaOH and ethyl acetate
Volume of solution in reactor: 2 ℓ

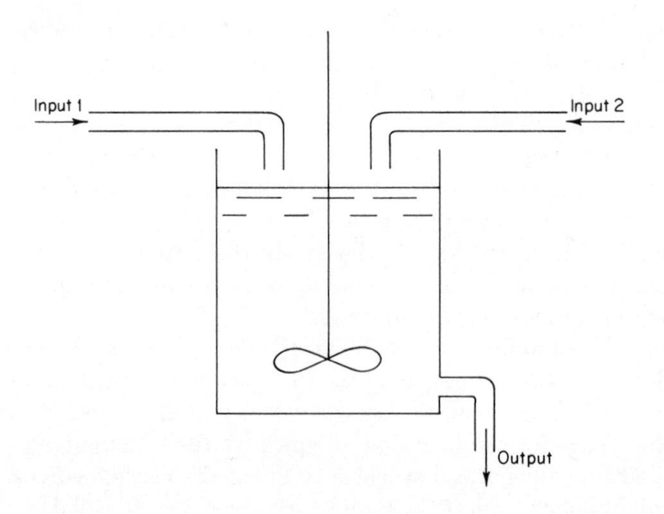

Fig. 7.5 A well-stirred reactor.

Then we can write, for either NaOH or ester,

rate of inflow $= 0.001 \times 0.1 = 10^{-4}$ mole/second
rate of outflow $= 0.002 \times 0.02 = 0.4 \times 10^{-4}$ mole/second
rate of reaction $\times$ volume (assuming a second-order rate constant k) $= 2k(0.02)(0.02) = 8 \times 10^{-4}\, k$ mole/second

For a material balance

$$10^{-4} - 8 \times 10^{-4}k = 0.4 \times 10^{-4}$$

or

$$k = 0.075 \quad \text{mole}^{-1}\, \ell\, \text{sec}^{-1}$$

While well-stirred reactors are most effective for processes that occur at a leisurely pace, with efficient stirring they can be used with average residence times of a minute or less for the study of reactions that could hardly be approached in a conventional static reactor.

7.4 FLOW METHODS FOR FAST REACTIONS

7.4.1 Tubular Flow Reactiors

The methods described above can be used to study reactions that are quite fast by increasing the flow rate and/or lowering the concentration. For example, as you calculated in Problem 2.10, the half-life of a second-order reaction 2A $\rightarrow$ products is given by

$$t_{1/2} = \frac{k}{[A]}$$

A stream of gas can flow at speeds of $\frac{1}{10}$ or more the speed of sound (5000 cm/sec would be quite possible) while at a pressure of 0.01 Torr (5×10^{-7} mole/ℓ near room temperature). If the reactor is 50 cm long, then a rate constant of 2×10^8 mole ℓ^{-1} sec^{-1} could easily be measured. Under these conditions, flow would be molecular and the calculated residence time of 0.01 sec would be realistic.

This approach has been used to obtain rate constants for a number of gaseous bimolecular reactions, such as $O + H_2 \rightarrow OH + O$ and $O + CH_4 \rightarrow OH + CH_3$. The radicals or atoms are produced in a side tube in an electrical discharge, then introduced into the main stream of the reactor, which contains the reactant molecule. Reaction times may be varied by changing both the flow speed and the distance of the observation point from the mixing location. Typically, some spectroscopic method of analysis is used; in the examples quoted, the free radicals were determined by electron spin resonance (18).

Fast flow reactors were the first type used to study rapid reactions in solution (19), partly because fast analytical methods were less available in the 1920s they are now. The continuous flow approach has gone out of fashion now because large volumes of solution were required by the time the flow rates were adjusted and measurements taken by conventional methods. One variation that is still used involves a continuous flow for a short period of time, during which the flows are increased steadily (and rapidly) so that a detector at a fixed location observes solutions that have been mixed for decreasing times. Since the whole process can be carried out in less than a second, the volumes of reactants are not unreasonable (20).

7.4.2 Shock Tube

The shock tube is an instrument that can function as either a flow or static reactor in gas kinetics. As illustrated in Fig. 7.6, a shock tube consists basically of a length of pipe (typically 3–10 cm in diameter and 2–6 m long) divided in the middle by a rupture disk, or diaphragm. The expansion tank is used only for static experiments in which gas samples need to be taken for analysis outside the shock tube. In a shock tube a sample of gas can be heated very rapidly (in terms of one to ten microseconds for a 1000° change, depending on the speed at which equilibration of energy among the degrees of freedom occurs), held at high temperature for several milliseconds, then cooled at a rate of close to a million degrees

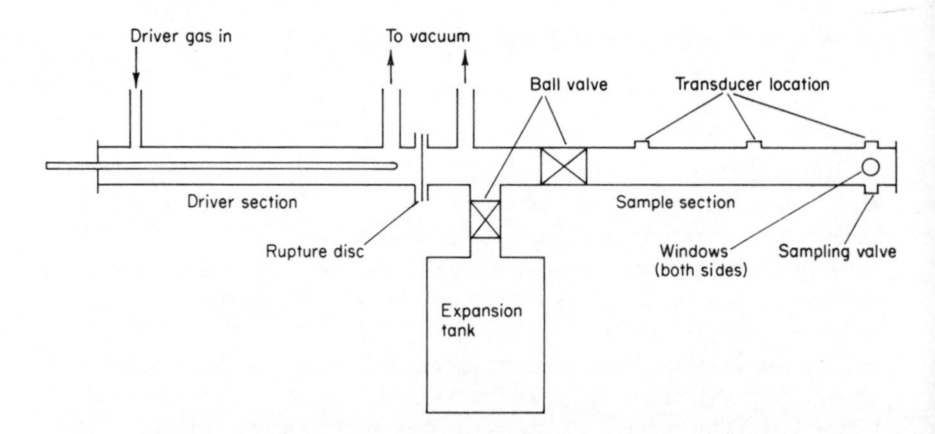

Fig. 7.6 Schematic drawing of a shock tube.

per second, and this is done without contact of the gas with a hot wall (which could be catalytic) and without the use of radiation (which could cause specific energy changes in the molecules, as in flash photolysis.)

In a typical experiment in which the expansion tank would be used, a high pressure (10 atm, for example) of helium is put into the driver section, and a low pressure (~0.2 atm) into the sample section and the expansion tank, the valve to the tank being open. When the rupture disk is broken either by the helium pressure or by the plunger, the tube suddenly has a very big step in pressure in the middle, with nothing to maintain it. What follows is the interesting phenomenon called a shock wave, which can best be described by a series of diagrams (Fig. 7.7). In these, pressure is plotted against distance along the shock tube, at a series of times which are typical for shock tube operation.

Figure 7.7a shows the original driver and sample gas pressures just before the rupture disk breaks. After a millisecond (Fig. 7.7b) the shock wave (sudden jump in pressure of the sample gas) is well developed and is propagating down the tube at about 1000 m/sec. To the left of the shock wave is a region of constant pressure which includes not only the part of the sample that has been "shocked," but also some of the helium that has expanded. This gas is moving to the left at a speed of about 700 m/sec. The sample is heated by the sudden compression, while the driver gas is cooled by expansion. The dotted line represents the interface between sample and driver gas; this region may be a turbulent zone because of the finite time it takes to break the diaphragm. To the left of the region of uniform pressure, the driver gas is expanding, which still farther to the left it is at its original pressure, since the expansion wave propagates back into the driver gas at the speed of sound (about 1000 m/sec for helium at room temperature). At the far right is a region of undisturbed sample gas, still at its original pressure and temperature, and "unaware" that anything has happened in the shock tube. It will be clear from this description that the process occurring here is quite different from that occurring, for example, in the cylinder of an automobile engine. In the latter case compression occurs slowly enough that the entire gas sample is at nearly the same pressure

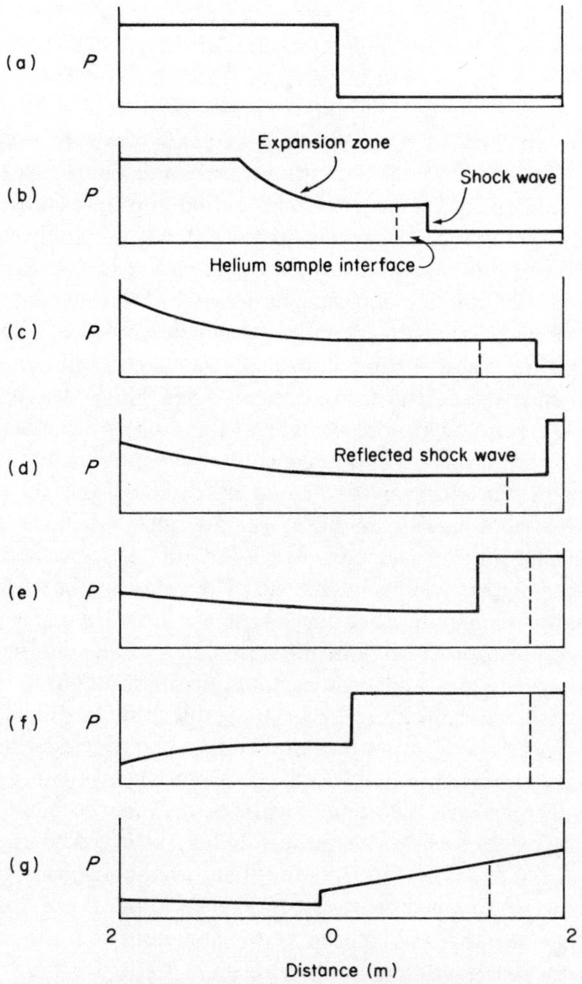

Fig. 7.7 Graphs showing typical pressure changes in a shock tube.

throughout the process, and from the thermodynamic point of view the compression is nearly reversible and isentropic. In the shock process the compression is highly irreversible, and it can be shown that the entropy of the sample gas increases as it is traversed by the shock wave. Accordingly, the temperature rise for a given pressure rise in a shock wave is greater than in an isentropic compression. Besides that, the shocked gas sample has a substantial amount of directed translational energy along the tube; this energy can be converted to thermal energy if the gas is brought to rest.

We can see that in Fig. 7.7c the shock wave has traversed most of the sample gas, heating it and setting it in motion to the right. Observations of this hot, flowing gas can be made spectroscopically through windows set in the tube wall, the gas pressure being measured directly by the fast-response pressure transducers shown, and the temperature and flow rate calculated from the pressure and from the speed of the shock wave (obtained by noting the time between the arrival of the shock wave at two transducers a known distance apart). We should note that the pressure and temperature behind a shock wave remain nearly constant over a period of milliseconds. That is, this shock wave is a boundary between gas in two states, one cold (stationary and at low pressure) and the other hot (moving and at high pressure). We do not study reaction kinetics *in* the shock wave, but *behind* it.

Figure 7.7d shows what happens when the shock wave has traversed all of the sample gas. In a closed tube, flow of the gas to the right cannot continue for long, and we see the formation of a new shock wave, called a "reflected shock wave," that corresponds to the stopping of the gas flow by the end plate of the tube. The momentum of the moving gas leads to a higher pressure in the stopped gas, while its directed kinetic energy is converted into thermal energy. In Fig. 7.7e, the reflected shock wave has traversed all of the sample and part of the helium. Observations can be made on the sample as it resides in the end of the shock tube at high temperature. Probably the commonest measurements have been spectroscopic (both emission and absorption); but measurements of gas density, rate of heat transfer from the gas, and even removal of small amounts of gas for mass spectrometric analysis during the few milliseconds available have been made. In some ways, observations behind the reflected shock wave are simpler than those behind the incident wave because the gas is stationary in the former case, while observations behind the incident wave must allow for the fact that the sample being observed is continually changing due to the gas flow. On the other hand, temperatures behind the reflected wave tend to be less certainly known than those behind the incident wave because there have been more opportunities for nonidealities of gas flow (particularly effects due to viscosity) to affect them.

So far we have neglected the presence of the expansion tank. While the helium is flowing rapidly along the tube, it tends to bypass the tank, but when the reflected shock wave brings the helium to rest, flow into the tank occurs, so that the pressure in the shock tube drops rapidly Fig. 7.7g. In particular, the expansion of the sample causes rapid cooling and quenching of reaction, so that part of the sample can now be taken for a relatively leisurely analysis outside the shock tube.

To summarize, shock tubes can be used as either flow or static reactors to study gas reactions at high temperatures. Measurements can be made on the gas samples either during the heating cycle of afterward. Shock waves simply result in rapid temperature and pressure changes in the gas, no special effects (such as photochemical excitation) being produced. Several free-radical reactions, such as the pyrolysis of methane and ethane, and the oxidation of hydrogen, which were discussed in Chapter 4, have been studied in shock tubes. Also, a number of unimolecular reactions, such as the cis–trans isomerization of butene-2, which can be analyzed by the RRKM theory, have been studied. A typical shock-tube study, carried out by the author on the reactions of ethylene at high temperature, is described in Ref. 21. It should be emphasized that, as with other types of apparatus, the particular design of a shock tube will depend on the experiments that are planned for it, and to some extent, the background and prejudices of the experimenter. The sketch in Fig. 7.6, which is based of the shock tube of Ref. 21, is simply one of the several designs that have been found to work.

7.4.3 Molecular Beams

A development comparable in importance to the relaxation methods has been the appearance of molecular beam and mass spectrometric techniques for the study of individual molecules as they react. These methods have allowed experimenters to come closer than ever before to the study of Johnston's "chemical–physical" reactions, in which the details of the interactions of individual molecules, rather than the average result of large numbers, can be studied. Actual results from these methods have been less numerous than those from relaxation methods, since the techniques are very difficult, but rapid progress is being made.

A molecular beam consists of a stream of molecules, typically at low pressures so that few, if any, intermolecular collisions occur within the beam. Such a beam can be produced only in a high vacuum, where other molecules will not disturb it. The source of molecules is typically a container with a small hole, through which molecules emerge at all angles, while a beam is produced by selecting only those molecules which emerge at a particular direction (Fig. 7.8a). Such a molecular beam will contain molecules with a thermal distribution of energy. Their average translational

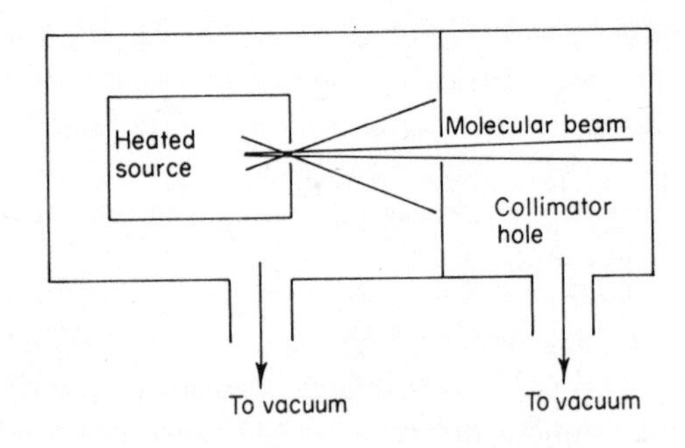

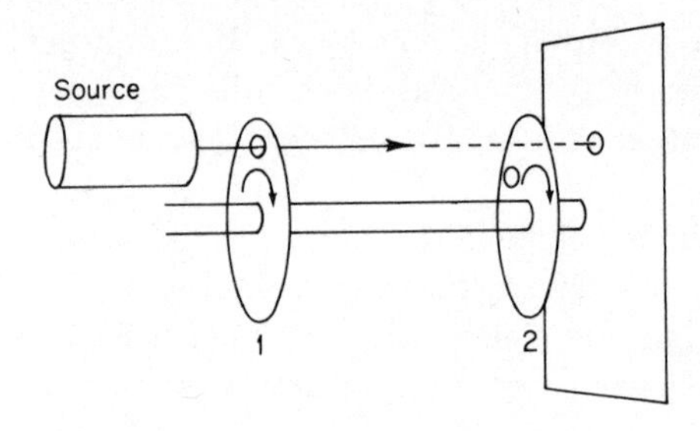

Fig. 7.8 Molecular beam apparatus. (a) Formation of a molecular beam; (b) velocity selector.

energy will correspond to the source temperature (being $\frac{3}{2}RT$), while they will have the rotational and vibrational energy distribution appropriate to the source temperature.

It is possible to select those molecules that have a small range of translational energies (that is, speeds along the direction of the beam) by placing an apparatus like that of Fig. 7.8b between the source and the collimator hole. A group of molecules with the full range of thermal velocities will pass through the hole in plate 1 when it is opposite the source. However, only those with the desired velocity will reach plate 2 when its hole is opposite the collimator hole. The efficiency of the device is increased if several holes are placed in each plate, but it is clear that, by "throwing away" the molecules that have velocities outside a narrow range, the intensity of the beam is much reduced in the selection process. The problem of selecting molecules with a given amount of rotational energy is more difficult, and not much has been accomplished in this direction, although some success has been obtained by exciting molecules to higher-than-thermal rotational energies by absorption of radiation. Fortunately, for many molecules there is no problem with the vibrational energy distribution—nearly all the molecules are in the lowest vibrational state at moderate temperatures.

It is easy to see in principle, then, how two such molecular beams can be arranged to intersect, or cross, one another, so that the molecules will collide in a controlled way. Since the beams are dilute (to avoid intrabeam collisions), only a small fraction of molecules of each type will collide; but when they do the reaction products will be scattered away from the original directions of the beams, so that to follow the reaction we would want a detector that could (a) be movable to different angles from the crossing-point of the beams, to enable us to study the distribution of the scattered molecules, and (b) be able to distinguish among the reactants and products, determine their numbers, and ideally measure their energies also. This is a big order, and actual detectors only approximate these desired qualities.

What would happen if we carried out a molecular beam experiment with two given reactants, gradually increasing the energies of the reacting molecules? Three kinds of intermolecular processes will be found. At quite low translational energies, the colliding molecules will act strictly as elastic bodies, bouncing off one another in a mechanical way just as two billiard balls, or two argon molecules, would do. Since the molecular masses and initial velocities are known and since molecules undergoing elastic collisions tend to act as hard spheres even though their actual shapes are not quite spherical, it is possible from the measured distribution of scattered molecules to calculate an effective radius (or cross section) for each type of molecule for elastic scattering. In general, these dimensions will be close to those determined by kinetic-theory methods.

At somewhat higher energies (but still below the activation energy) a new kind of collision will appear in which one or both of the molecules gains or loses some internal energy. For example, if an atom struck a diatomic molecule in the right way, it could excite the molecule from the rotational state with $J = 2$ to that with $J = 3$. Translational energy will disappear in the process, as will translational momentum, since the angular momentum has increased, so that the simple scattering pattern that existed before is changed. Collisions of this type are called *inelastic* since the molecules do not act as hard, perfectly elastic spheres in them, but are able to absorb or give up internal energy as well.

At still higher energies, the availability of the activation energy *may* lead to reaction if the collision occurs in just the right way. For these reactive collisions, the angular distribution of reaction products can be very informative, since we can visualize many ways in which a reaction could occur. For example, an atom could collide with a molecule, knocking off part of the molecule and carrying it along, so that one of the reaction products tends to be found in the general direction of the initial beam of atoms. Conversely, a relatively long-lived complex could form, which could rotate several times and then, on decomposition, send reaction products to all angles nearly equally. In every case there must be overall conservation of mass, momentum, and energy, but there are many possibilities within these limitations. The approach is to determine as much as possible about the angular distributions, then deduce how these could have been produced. Of course, at a given energy all three kinds of collisions—elastic, inelastic, and reactive—may occur, depending on how the collisions occur. Glancing collisions will tend to be unreactive unless the energy is very high compared to the activation energy, while the likelihood of reaction in nearly head-on collisions will depend on the specific orientation of the molecules to one another at the moment of collision.

A small but fascinating subdiscipline of kinetics has sprung up within the past ten years, consisting of the simulation of these molecular interactions by computer. Assumptions are made as to the potential energy changes that occur as the molecules approach one another, initial velocities and internal states of the molecules are set, and then the molecules are (mathematically) started toward one another along trajectories that will bring them into collisions that can vary from head-on to merely a slight change of path due to intermolecular attractions. Such calculations are just what is needed to interpret molecular beam results, the calculations becoming more accurate and realistic as they are modified to reproduce experiment.

One of the limitations of the technique described so far is the small amount of energy available from thermal translational motion. An energy of $\frac{3}{2}RT$ is only 1 kcal/mole at room temperature, or 3 kcal at 1000° K. Accordingly the first molecular beam experiments involved reactions with very low activation energies, such as (22)

$$K + HBr \rightarrow H + KBr$$

(another reason for studying this reaction was that there is a very sensitive method for detecting K atoms). To go to higher energies, the mass spectrometric method of accelerating positive ions with an electric field has been used, and the area of ion–molecule reactions has developed rapidly in recent years. Since acceleration of an ion by a field on 1 V gives it an energy equivalent of 23 kcal/mole, there is no problem of getting to high energies by this method, while analysis of ionic products of reaction by mass-spectroscopic methods is a well-established technique. For example, the speed of such ionic products can be measured by applying a retarding field at the entrance to the detector. While some of the ions worked with seem rather strange to a kineticist working in other areas, the mass spectroscopist will argue with conviction that his results obtained with the Ar^+ ion will be similar to those to be expected with the Cl atom (since they are isoelectronic) with some changes due to the presence of the charge, which tends to increase intermolecular attractions and hence the rate constants.

References

1. F. W. Schneider and B. S. Rabinovitch, *J. Amer. Chem. Soc.* **84,** 4215 (1962).
2. L. S. Kassel, *J. Amer. Chem. Soc.* **54,** 3949 (1932).
3. G. B. Skinner and R. A. Ruehrwein, *J. Phys. Chem.* **63,** 1736 (1959).
4. G. W. Hoffman, *Rev. Sci. Instrum.* **42,** 1643 (1971).
5. M. Eigen and J. Schoen, *Z. Electrochem.* **59,** 483 (1955); M. Eigen and L. De Maeyer, *Z. Electrochem.* **59,** 986 (1955); M. Eigen, W. Kruse, G. Maass, and L. De Maeyer, *Progr. React. Kinet.* **2,** 285 (1964); B. R. Staples, D. J. Turner, and G. Atkinson, *Chem. Instrum.* **2,** 127 (1969); J. J. Auborn, P. Warrick, Jr., and E. M. Eyring, *J. Phys. Chem.* **75,** 2488 (1971).
6. C. N. McDowell, D. O. Hansen, and N. L. Roy, *Rev. Sci. Instrum.* **42,** 163 (1971).
7. L. Onsager, *J. Chem. Phys.* **2,** 599 (1934).
8. M. Eigen and K. Tamm, *Z. Elektrochem.* **66,** 93, 107 (1962).
9. P. Hemmes and S. Petrucci, *J. Phys. Chem.* **72,** 3986 (1968); **74,** 467 (1970).
10. H. S. Gutowsky and A. Saika, *J. Chem. Phys.* **21,** 1688 (1953).
11. T. J. Pinnavaia and A. L. Lott, II, *Inorg. Chem.* **10,** 1388 (1971).
12. B. G. Willis, J. A. Bittkofer, H. L. Pardue and D. W. Margerum, Anal. Chem. **42,** 1340 (1970).
13. G. Porter and F. J. Wright, *Disc. Faraday Soc.* **14,** 23 (1953); G. Porter, *Science* **160,** 1299 (1968).
14. J. E. Nicholas and R. G. W. Norrish, *Proc. Roy. Soc. Ser. A* **307,** 391 (1968).
15. K. H. L. Erhard and R. G. W. Norrish, *Proc. Roy. Soc. Ser. A* **259,** 297 (1960).
16. G. Porter and M. R. Topp, *Proc. Roy. Soc. Ser. A* **315,** 163 (1970).
17. R. V. Poirier and R. W. Carr, Jr., *J. Phys. Chem.* **75,** 1593 (1971).
18. A. A. Westenberg, *Science* **164,** 381 (1969); A. A. Westenberg and N. de Haas, *J. Chem. Phys.* **50,** 2512 (1969).
19. H. Hartridge and F. J. W. Roughton, *Proc. Roy. Soc. Ser. A* **104,** 376 (1923).
20. B. Chance, *Rev. Sci. Instrum.* **22,** 619, 627, 634 (1951).
21. G. B. Skinner, R. C. Sweet, and S. K. Davis, *J. Phys. Chem.* **75,** 1 (1971).
22. E. H. Taylor and S. Datz, *J. Chem. Phys.* **23,** 1711 (1955).

Further Reading

P. G. Ashmore, F. S. Dainton, and T. M. Sugden, eds., "Photochemistry and Reaction Kinetics." Cambridge Univ. Press, London and New York, 1967.

F. E. Caldin, "Fast Reactions in Solution." Blackwell, Oxford, 1964.

S. Claesson, ed., "Fast Reactions and Primary Processes in Chemical Kinetics." Wiley (Interscience), New York, 1967.

S. L. Friess, E. S. Lewis, and A. Weissberger, eds., "Technique of Organic Chemistry, Vol. VIII, Investigation of Rates and Mechanisms of Reactions, Parts I and II," 2d Ed. Wiley (Interscience), 1961 and 1963.

A. G. Gaydon and I. R. Hurle, "The Shock Tube in High-Temperature Chemical Physics." Reinhold, New York, 1963.

E. W. McDaniel, V. Cermak, A. Dalgarno, E. E. Ferguson, and L. Friedman, "Ion-Molecule Reactions." Wiley (Interscience), New York, 1970.

H. Melville and B. G. Gowenlock, "Experimental Methods in Gas Reactions," 2d Ed., Macmillan, London, 1964.

I. O. Sutherland, The investigation of the kinetics of conformational changes by nuclear magnetic resonance spectroscopy, in *Annual Reports on NMR Spectroscopy* **4**, 71 (1971).

Problems

7.1 (a) Place 250 ml of water in a 400 ml beaker. From appropriate measurements calculate the area of the beaker that is in contact with the solution.

(b) Suppose that this beaker of water, at 25° C, is placed in a large thermostat bath filled with water at 50° C. Given that the thermal conductivity of the glass is 0.003 cal cm^{-1} sec^{-1} K^{-1} and that the beaker wall is 2 mm thick, obtain a graph showing the temperature of the water in the beaker as a function of time. Feel free to make simplifying assumptions such as: the specific heat of the glass is negligible compared to the water; the water both in and out of the beaker is well stirred so the temperature gradient is entirely across the glass; and so on. Note the formal similarity of your solution to that for a first-order chemical reaction.

(c) Make a comparable graph using the same basic assumptions, but for 250 ml of air at 1 atm pressure (initially) with the same surface area and the same initial temperatures.

7.2 In the flash photolysis experiment described in the text, it was stated that "flashing" a mixture of Cl_2 and O_2 produced Cl atoms, but nothing was said about O atoms. If it is assumed that the light used has wavelengths from 3000 to 7000 Å, show why one would expect to obtain Cl, but not O.

7.3 (a) Suppose we had a tubular reactor 30 cm long and 0.4 cm internal diameter. For an aqueous solution near room temperature, what will be the shortest average dwell time in the tube if the Reynolds number is not to exceed 1000?

(b) Calculate the pressure needed to make the fluid flow at that rate.

(c) Calculate the velocity profile across the tube, assuming the flow is laminar.

(d) Make comparable calculations for air as fluid, assuming that it should emerge from the tube at 1 atm. Does the compressibility of air make a significant difference in your calculation?

7.4 While, in describing the "well-stirred reactor," we talked about measuring the input and output streams, we could also have considered measuring the concentrations *in* the reactor since presumably these are the same as in the output stream. Suppose we put a stirrer in a heated infrared gas cell with a total volume of 100 cc, and flowed in a gas mixture containing 5% of a reactant that had a first-order rate constant for conversion to a product of 0.1 sec^{-1}. Calculate the concentration of A in the infrared cell for flow rates of 0.3, 1, 3, 10, and 30 cc/sec, measured at the cell temperature. Approximately how rapidly must the gas be heated so that less than 1% would react during the heating time?

7.5 We mentioned that when observations are made on gas flowing in a shock tube we must "allow for the fact that the sample being observed is continually changing due to the gas flow." Suppose, as in the text, that the speed of the shock wave is 1000 m/sec, while the gas flow speed behind the wave is 700 m/sec. If the shock wave passes an observation station at time t_0, and a gas sample is observed at the station 200 microsec later, calculate the time interval that passed between the heating of the sample by the shock wave and the time of the observation. Show that the ratio of the second time to the first would be the same for other experimental times with the same shock and flow speeds. This ratio is sometimes called the ratio of "particle time" to "laboratory time." It would, of course, vary from one experiment to another.

7.6 Look up one or more recent journal articles in which kinetic results are described. Evaluate the author's choice of method, the special adaptions necessary to obtain the data, and the difficulties encountered. As appropriate, suggest alternative methods or minor variations in technique that might lead to better results. Finally, are there other measurements that could be made to answer questions that are left unanswered in the article?

Appendix

Some Important Thermodynamic Relationships

A. COMMONLY USED SYMBOLS IN THERMODYNAMICS

C	Specific heat	P	Pressure
E	Internal energy	Q	Partition function
$\mathcal{E}$	Electrical potential	S	Entropy
G	Gibbs free energy	T	Temperature
H	Enthalpy or heat content	V	Volume
K	Equilibrium constant		

These symbols are the ones used in this text, and are widely used in the chemical literature. Students should be aware, though, that they are not universally used by convention. For example, physicists often use U, rather then E, for internal energy.

When it is desired to express the change in a thermodynamic quantity that occurs in a physical or chemical change, the symbol Δ is put in front of the thermodynamic symbol: for example, ΔE signifies the change in E.

Normally, the thermodynamic quantities are given in terms of one mole of a substance. For a change, they will refer to the chemical equation as written, taken in terms of moles.

A superscript degree sign ° signifies that the quantity refers to the substance in its standard state.

Subscripts are used for several purposes. When thermodynamic quantities are used as subscripts, the usual meaning is that these quantities are kept constant. For example, C_P signifies "the specific heat at constant pressure." Temperatures are often given as subscripts. For example, $H_{298} - H_0$ signifies "the enthalpy at 298°K minus the enthalpy at 0°K." Subscripts are sometimes used to denote the variables under two different sets of conditions that are defined somewhere else. For example, $H_2 - H_1$ could signify "the enthalpy in state 2 minus the enthalpy in state 1." Clearly, the last two uses of subscripts will have to be distinguished by context.

B. THERMODYNAMIC RELATIONSHIPS

$$H = E + PV$$

$$H = E + RT \qquad \text{only for a mole of an ideal gas}$$

$$C_V = \left(\frac{\partial E}{\partial T}\right)_V$$

$$C_P = \left(\frac{\partial H}{\partial T}\right)_P$$

$$S_2 - S_1 = \int_{T_1}^{T_2} \frac{C_p\,dT}{T} = \int_{T_1}^{T_2} C_p\,d(\ln T)$$

$$G_2 - G_1 = RT \ln \frac{P_2}{P_1} \qquad \text{only for a mole of an ideal gas}$$

$$\Delta G^\circ = -RT \ln K$$

$$\Delta G = -n\mathfrak{F}\mathcal{E}$$

where n is the number of moles of electrons involved in the electrochemical reaction.

Answers to Selected Problems

Chapter 2

2.1 Rate constants: (a) 7×10^{10} mole^{-1} cc sec^{-1}; (b) 7×10^{4} mole^{-1} m^{3} sec^{-1}; (c) 1.17×10^{-13} molecule^{-1} cc sec^{-1}; (d) 4.2×10^{12} mole^{-1} cc min^{-1}
Rates: 4.2×10^{-5} mole ℓ^{-1} sec^{-1}; (a) 4.2×10^{-8} mole cc^{-1} sec^{-1}; (b) 4.2×10^{-2} mole m^{-3} sec^{-1}; (c) 2.52×10^{16} molecule cc^{-1} sec^{-1}; (d) 2.52×10^{-6} mole cc^{-1} min^{-1}.

2.2 Rate $= 1.70 \times 10^{-11}$ mole ℓ^{-1} sec^{-1}; $d[\mathrm{I}]/dt = -3.4 \times 10^{-11}$ mole ℓ^{-1} sec^{-1}; $d[\mathrm{HI}]/dt = 3.4 \times 10^{-11}$ mole ℓ^{-1} sec^{-1}; $d[\mathrm{H_2}]/di = -1.7 \times 10^{-11}$ mole ℓ^{-1} sec^{-1}.

2.5 From kinetic data, $K = 0.14$; from thermodynamic data, $K = 0.26$.

2.6 For a second-order reaction for 2 $CH_3OC_6H_4NO \rightarrow$ product, $k = 0.0074$ mole^{-1} ℓ min^{-1}.

2.7 Order re F^- is close to 1. Overall second-order rate constant is 0.29 mole^{-1} ℓ sec^{-1}.

2.8 $k = 0.0294$; doubling time = 24 years; extrapolation to 1970 gives 770,000,000.

2.12 For example, at time $= 10^5$ sec, at constant P, $[\mathrm{A}] = [\mathrm{A}]_0/(2e - 1)$ and $V = V_0(2e - 1)/e$. At constant V, $[\mathrm{A}] = [\mathrm{A}]_0/e$.

2.13 $k = 6.3 \times 10^7\, e^{-5260/RT}$, R and E in calories.

2.14 A, 26.6 kcal.; B, 67.8 kcal; C, 29.2 kcal; D, 9.9 kcal.

2.15 (a) $K_c = 3680\, e^{-103{,}900/RT}$; (b) $k_- = 5.5 \times 10^7\, e^{8{,}300/RT}$; (c) $\Delta S^\circ = 29.2$ cal deg^{-1}, $\Delta H^\circ = 109{,}900$ cal; (d) $\Delta S^\circ = 29.26$ cal deg^{-1}, $\Delta H^\circ = 109{,}840$ cal.

Chapter 3

3.3 0.85; 0.19; 6×10^{-8}; 5.3×10^{28}; 1.2×10^{28}; 3.7×10^{21}.

3.4 Calculated σ_R values: 7.7 Å for HO_2 + Ar; 0.12 Å for CO + OH.

3.5 504 cm^{-1}; 6×10^{10} erg/mole; 6,000 J/mole; 1,440 cal/mole. Ratios 0.00061, 0.23, 0.48, 0.78 to 1.

3.6 Equilibrium constants: 0°, 0.000; 100°, 0.040; 1000°, 1.54; ∞ temperature, 2.00. Fraction of B molecules: 0°, 0.000; 100°, 0.038; 1000°, 0.61; ∞ temperature, 0.667.

3.9 (a) 4; (b) 3; (c) 4; (d) 2.

Chapter 4

4.1 $E_{CT} = 37{,}900$ cal; $k_2 = 2.4 \times 10^{13}$; $k = 7.5 \times 10^{-3}$; $k_{-1} = 1.94 \times 10^{14}$.

4.2 (a) $(5, 1, 0)$, $(2, 3, 0)$, $(3, 1, 1)$, $(1, 1, 2)$, $(0, 3, 1)$; (b) 28; (c) 108; (d) 5.8; (e) 1.94×10^{-4} if we take $N(E) = 0.085/cm^{-1}$.

4.7 (a) 9.3×10^{-12} mole/cc, 8.4×10^{-10} mole/cc;
(b) 5.2×10^{-11} mole/cc, 1.7×10^{-8} mole/cc; (c) 2.7×10^{-2}.

4.8 Rate $= k_3(k_1/k_5)^{1/2}[C_2H_6]^2$.

4.9 83, 68, 83, and 83 kcal.

4.10 0.064, 0.0065

4.11 7×10^{15} and 12×10^{15} $mole^{-2}$ cc^2 sec^{-1}.

Chapter 5

5.1 273 atm or $\approx$4000 psi.

5.2 (a) 3.2×10^4 cm/sec; (b) 1.6×10^{13} per second;
(c) 8×10^9 per second; 3.6×10^9 per second;
(e) 3.6×10^2 cm/sec.

5.3 (a) 1.2×10^{11} sec^{-1}, 40 cm^{-1}; (b) 5.2 cal deg^{-1} $mole^{-1}$;
(c) 9.8 cal deg^{-1} $mole^{-1}$, -5.0 cal deg^{-1} $mole^{-1}$.

5.4 (a) log $k°$ values are 1.43, 1.71, 2.05, 2.52, 3.04;
(b) $6.3 \times 10^{23}\, e^{-33{,}600/RT}$; (c) 33.0 kcal, 30.2 cal deg^{-1} $mole^{-1}$, 15.2 cal deg^{-1} $mole^{-1}$.

5.5 (a) 2.9×10^{-7}, 3.7×10^{-5}, 2.0×10^{-4}, 1.7×10^{-8}, 6.1×10^{-7}, 1.7×10^{-7} sec^{-1}; (b) 1.2×10^4, 1.5×10^8, 2.4×10^8; (c) 0.55, 0.68.

5.6 (a) -9.07; (b) 5.3×10^{-4}.

5.7 -0.7; -1.7.

5.9 0.0167 mole ℓ^{-1} sec^{-1}, 0.00257, mole ℓ^{-1}, 3×10^{-4} sec^{-1}, 0.00257 mole ℓ^{-1}.

5.10 $K_m = 0.00046$ mole ℓ^{-1}.
5.11 3.2×10^7 mole^{-1} ℓ sec^{-1}.
5.12 3×10^6 molecules.

Chapter 6

6.2 (a) 3.5×10^{-16} cm^2/sec; 910°C.
6.6 (a) 1.09×10^{-9} mole sec^{-1} cm^{-2}; (b) 1.05×10^{-6}, 0.46×10^{-6} mole sec^{-1} cm^{-2}; (c) 2.5×10^{-9} mole sec^{-1} cm^2; (d) fits case (a), $\theta_1 \rightarrow 1$.
6.10 Graph of log rate versus log P gives slope of n. Slope of 0.5 is reasonable if not too much weight is given to the first point.

Chapter 7

7.3 (a) 1.2 sec. (b) 1.85×10^{-4} atm pressure difference.
7.4 0.00146, 0.0045, 0.0115, 0.025, and 0.0375 mole % reactant; 0.1 sec.
7.5 667 μsec.

Author Index

Numbers in italics indicate the page on which the complete reference appears.

M

N

O

P

R

S

T

U, V

W, Y

Subject Index

I

J, K

L

M

N

O

P

Q, R

S

T

Physical Constants

The following material was taken chiefly from *J. Chem. Ed.* **48,** 569 (1971).

Avogadro's number,	N or N_A	6.022×10^{23} mole^{-1}
Boltzmann constant,	$\boldsymbol{k}$	1.38×10^{-16} erg deg^{-1} 1.38×10^{-23} J · K^{-1}
Gas constant,	$R = N\boldsymbol{k}$	8.314×10^{7} erg deg^{-1} mole^{-1} 8.314 J · K^{-1} · mole^{-1} 1.987 cal deg^{-1} mole^{-1} 82.1 cc-atm. deg^{-1} mole^{-1} 0.0821 l-atm deg^{-1} mole^{-1}
Electron charge,	e	4.803×10^{-10} esu 1.602×10^{-17} C
Faraday,	$\mathfrak{F}$	9.65×10^{5} C
Planck's constant,	h	6.626×10^{-27} erg sec 6.626×10^{-34} J · sec
Velocity of light,	c	2.998×10^{-10} cm sec^{-1} 2.998×10^{-8} m · sec^{-1}
Ratio of circumference to diameter of circle,	π	3.1416
Base of natural logarithms,	e	2.718

Conversion Factors

1 dyne of force	$= 10^{-5}$ N
1 angstrom	$= 10^{-10}$ m $=$ 0.1 nm
1 cm^{-1} (as energy unit)	$=$ 11.96 J · mole^{-1} $=$ 1.986×10^{-23} J · molecule^{-1} $=$ 1.986×10^{-16} erg molecule^{-1} $=$ 2.86 cal mole^{-1}

A B C D E F G H I J

Skinner, Gordon.

Introduction to chemical kinetics

DATE			